与 Bézier 对话

——常庚哲教授文选

常庚哲 ● 著

中国科学技术大学出版社

图书在版编目(CIP)数据

与 Bézier 对话:常庚哲教授文选/常庚哲著. —合肥:中国科学技术大学出版社,2015.10
ISBN 978-7-312-03854-9

Ⅰ. 与… Ⅱ. 常… Ⅲ. 计算机辅助设计—几何—造型设计—文集 Ⅳ. TP391.72-53

中国版本图书馆 CIP 数据核字(2015)第 244186 号

出版 中国科学技术大学出版社
     安徽省合肥市金寨路 96 号,230026
     http://press.ustc.edu.cn
印刷 合肥市宏基印刷有限公司
发行 中国科学技术大学出版社
经销 全国新华书店
开本 710 mm×1000 mm 1/16
印张 11
字数 192 千
版次 2015 年 10 月第 1 版
印次 2015 年 10 月第 1 次印刷
定价 30.00 元

序

  1954年夏天，我高中毕业. 高考前的体检，查出我患"色弱"，理工科各系，只有唯一的学科——数学可选. 于是我被分配到南开大学数学系. 现在想来，我没有其他的特长，只能够鼓捣鼓捣数学.

  1958年10月初的一个下午，阳光明媚. 南开大学数学系毕业生在芝琴楼召开分配派遣大会，几十个青年学子要在这里决定今后的命运. 全场鸦雀无声，大气也不敢出. 当总支书记念到"常庚哲，分配到北京中国科学技术大学"时，我真是喜出望外，不敢相信自己的耳朵，仿佛我是世界上最幸福的人. 我感谢南开大学数学系的师长，四年来，一直教育我、培养我，还送我到我梦寐以求、有着辉煌前途的名校. 当时中国科大刚刚成立，属于中国科学院，学生毕业后从事尖端科学技术研究，系主任都是各个领域的泰斗.

  10月14日下午，我从天津坐火车来到北京玉泉路报到，那一天蓝天白云、秋高气爽. 玉泉路大院的建筑，青砖灰瓦，没有高楼大厦，最多是二层楼房，安静而明亮. 我暗自问：何时能够见到华罗庚教授？

  起初，我跟随关肇直教授辅导学生四年，当他的助教. 这四年来，我跟关先生学习的很多东西是在南开大学不曾学习过的. 我要听关先生的课，要复习消化，要改学生的作业，要辅导学生，感觉力不从心. 后来在关先生指导之下，独立开课"几何与代数"两年. 接着在北京郊区两年"四清"，随之而来的是十年浩劫. 十二年的宝贵年华, 付之东流.

　　1972 年, 中国科大招收"工农兵学员", 读计算数学专业, 从整数、分数讲起. 1974 年, 这些学员来到贵州省安顺市镇宁县的云马机械厂进行毕业实习, 实习的课题是"飞机的机头罩和进气道的曲面外形", 用 Coons 曲面构造这些外形, 设备包括一台国产的电子计算机和日本产的数字绘图机. 尽管这一成果是初步的、原始的, 但具有划时代的意义. 北京航空航天大学著名的 CAD/CAM 专家唐荣锡教授, 在他的文章《CAD 产业发展的回顾与思考》中对此有生动的描述, 盛赞中国科大计算数学专业的学员们开创性的实践. 一次很偶然的"开门办学", 说来也很奇怪, 促成了我的毕生专业取向. 我从云马厂的航空工程师张金钟那里借来美国 AD 报告, 编号为 AD-663504, 就是后来著名的 Coons 曲面. 1977 年, 我发表长篇综述《Coons 曲面介绍》, 刊载于《计算机应用与应用数学》(1~24 页), 详细地描述了 Coons 曲面的理论和应用. 1979 年, 我和北京航空航天大学吴骏恒联合发表长篇综述文章《Bézier 曲线、曲面的数学基础及其计算》, 在《国外航空》杂志上连载 6 期.

　　我在西方专业杂志发表的第一篇论文, 就是《Mathematical Foundations of Bézier Technique》(与吴骏恒合作), 发表在英国 CAD 杂志上 (1981 年, 13 卷第 133~136 页). 在论文里我们提出了"Bézier 基函数"的名词, 证明了 Bernstein 基函数和 Bézier 基函数的相互关系. 过去我写的文章大都是综述性的, 没有科研成分可言. 接下来,《Matrix Formulations of Bézier Technique》, 也是刊载在 CAD 杂志上 (1982 年, 14 卷第 345~350 页), 我发现了 Bézier 基函数的一些性质. 后来, Bézier 教授与我有着通信联系, 但是我们两人不曾见面, 深为遗憾. 他送我两本书, 一本是法文的; 另一本是英文的, 并不厚, 书名叫《The Mathematical Basis of the UNISURF CAD System》, 其中提到了我的工作.

S. A. Coons（1912~1979）曾任美国麻省理工学院教授，P. Bézier（1910~1999）曾任法国雷诺汽车公司工程师，他们都是CAD的创始人和先驱者.

《Computer-Aided Geometric Design》（简称CAGD）杂志是1984年创刊的，我是第一任编委，而且是唯一的中国编委. 有人做过统计，在CAGD杂志的文章中，提到Bézier名字的占75%. 我将满80岁，细想起来，我的数学生涯，直接或间接与Bézier的名字有关联的，占绝大多数. 我在本书中写的15篇文章都是真实的故事，记载了我鼓捣数学的经验，有科研的、教学的、初等数学的，包括数学竞赛的. 只是年代久远，很多的时间记不得确切的日子.

北航（北京航空航天大学）吴骏恒教授会见Bézier总裁，那是1980年的事情. 雷诺汽车公司热情接待中国代表团一行，参观两周的时间. 后来，南航的丁秋林教授见过Bézier教授三次，其中有一次是他写了一本书，请Bézier作序，老先生欣然同意. 后来，丁秋林告诉我，在Bézier的客厅里，挂着一样东西，不是照片，不是名人字画，也不是中国山水，而是南京航空学院"请中国科大常庚哲副教授讲Bézier曲线、曲面"的海报. 老丁马上拍了下来，带回给我. 字迹和日期拍得清清楚楚，我珍藏至今. 老丁还拍了Bézier的单人相，老先生西装革履，个子很高，腰板笔挺，一派绅士风度.

提起我们的书名，邓建松教授有一个很好的建议：与Bézier对话，非常真实，又很浪漫，概括了我鼓捣数学的大致经历. 在这本书中，有一节叫"国际数学奥林匹克竞赛"，表面上没有Bézier的名字，实际上是从Bézier网构思出来的. 严格地说，在本书中Bézier的名字无所不在！

我的数学生涯离不开中国科大，离不开科大数学系，也离不开我们的CAGD小组. 20世纪80年代，国家公派冯玉瑜（1979~1981年）和我（1980~1982年）到美国当访问学者. 我们学成归国后，一起带硕士研究生和博士研究生，建立 CAGD 小组，开创研究方向，使队伍发展壮大. 随着我和冯玉瑜相继退休，近年来中国科大的 CAGD 研究小组队伍大大年轻化，形成了以陈发来、邓建松、刘利刚教授为主的年轻学术梯队. 研究小组成员之间胸怀坦荡，亲密无间，相互协作，共同进步，在有理曲线和曲面的隐式化、有理曲线和曲面的 μ 基理论、T 网格上的样条理论及等几何分析、分片代数曲面造型、隐式曲面重建、稀疏优化、几何处理及三维打印等方面的工作得到了国内外同行的关注与认可. 我们的研究小组自建立以来，培养了一大批硕士生、博士生. 他们中有些进入高等院校继续从事 CAGD 的研究，另一些则进入工业应用领域，都在各自的岗位上发挥了积极的作用. 衷心希望我们的 CAGD 研究小组继续人才辈出，兴旺发达！

常庚哲
2015 年 9 月

CONTENTS 目　次

序	i
1　回忆关肇直先生	1
2　大师风范	5
3　奇才怪杰　良师益友——忆曾肯成先生	7
4　三次 Hermite 多项式的基底	10
关于三次样条函数的两点注记	11
5　Pedoe 不等式	25
Proving Pedoe's Inequality by Complex Number Computation	29
6　Douglas-Neumann 定理	31
A Proof of a Theorem of Douglas and Neumann by Circulant Matrices	33
7　三角形曲面	37
The Convexity of Bernstein Polynomials over Triangles	41
8　保凸性定理	60
An Improved Condition for the Convexity of Bernstein-Bézier Surfaces over Triangles	63

| 9 | 第一次国际会议 | 70 |

Convergence of Bézier Triangular Nets and a Theorem of Pólya 74

| 10 | **Kelisky-Rivlin** 定理 | 88 |

Limit of Iterates for Bernstein Polynomials Defined on Higher Dimensional Domains 91

| 11 | 灯火阑珊觅丽人 | 97 |

Criteria for Copositive Matrices Using Simplices and Barycentric Coordinates 100

| 12 | 国际数学奥林匹克竞赛 | 123 |

| 13 | **Whitney** 定理 | 133 |

Problems and Solutions 135

| 14 | 命题五则 | 139 |

| 15 | 笛卡儿之梦 | 148 |

| 16 | **Over and Over Again** | 150 |

| 17 | 数学分析教程 | 153 |

| 18 | 重访 BYU | 157 |

附录　常庚哲简历 160

　　1　个人信息 160
　　2　获奖情况 161
　　3　学术任职 161
　　4　国际学术交流 161
　　5　出版著作 162
　　6　研究论文和综合报告 164
　　7　在国外数学期刊上提出的问题 167

回忆关肇直先生[①]

  1958年10月，我从南开大学毕业，被分配到北京的中国科学技术大学数学教研室. 龚昇同志接待了我，安排我到原子核物理和原子核工程系担任助教. 当年中国科大刚开始创办，开学已经快一个月了. 数学课的主讲教师是关肇直教授. 原子核物理和原子核工程系的新生近二百人，分成五个班. 周云龙和李端瑜各带两个班；我刚刚来到中国科大，他们发扬风格，让我带一个班.

  关先生是中国科学院数学研究所的研究员，学部委员，也就是现在的中国科学院院士. 关先生还担任研究所里的领导职务，在中国科大，除了繁重的教学工作，还兼任数学教研室主任. 那时，数学系和数学教研室是平级的单位. 关先生每周三半天在中国科大上班，从中关村到玉泉路有中国科大的班车，在中国科大食堂用早餐和午餐，中午搭班车回中关村. 中国科学院的研究员来中国科大上课，没有任何酬金，他们都看成自己分内的事情，唯有在中国科大上课的当天，两餐由中国科大食堂供应. 关先生的时间抓得很紧，除了上课，他总是在办公室工作. 他向助教们了解学生的情况，谈一些关于教学的事情. 我们尽量不打搅他，说完事也就走了. 他有一个很大的公文包，里面装满了书籍，沉甸甸的. 在他的中国科大办公室里，他一有空总是聚精会神地写讲义.

  [①] 节选自《中国科学技术大学数学五十年》，第117~119页.

1958 年招来的数学系首届新生，由华罗庚教授主讲数学课. 在基础课阶段，由华先生总揽全局，"一统天下"，这就是著名的"华龙"的概念. 1959 年，另外一条"龙"开始腾飞，即所谓的"关龙"；1960 年，又有吴文俊教授的"吴龙"紧随其后. 这三条"巨龙"，成为中国科大数学系的亮丽风景线. 这三位大专家的出发点和目标只有一个，那就是"多出人才，快出人才". 他们的经历、专业和兴趣各不相同，形成了相得益彰、百花竞放的生动局面.

关先生告别 1958 级原子核物理和原子核工程系的学生，1959 年秋天开始为数学系 1959 级学生上课，我也跟着关先生来到数学系. 说实在的，我非常感激他的培养、关心和信任. 在 1959 年，先后担任辅导教师的有范先信、黄开鉴和我.

数学系 1959 级的学生到学校里的时候，关先生写的新书《高等数学教程》第一册还没有正式出版，准备在高等教育出版社出书，关先生要我校对校样. 他对我说："由于我不拿稿酬，你的校对工作只能是义务性的，委屈你了. 不过，我会在前言之中，对你的劳动表示感谢."为了关先生出书，我经常跑高等教育出版社，该书的责任编辑杨锡文同我比较熟悉. 有一天，她和我谈起："关先生的《泛函分析讲义》也是我们社里出的，他不要稿酬，至今由我们社里暂存着."对关先生淡泊名利高风亮节的钦佩，溢于言表.

关先生的《高等数学教程》，总共出了三册. 第一册和第二册大致包括数学分析的内容，第三册是复变函数. 华先生的《高等数学引论》也出了三册，以后再也没有问世. 我想，这与我们国家当时的政治气候不无关系.

关先生请来数学研究所的丁夏畦先生为 1959 级学生上习题课，为关先生的正课配套. 后来，丁先生当选为中国科学院的院士. 记得丁先生为学生上"数学归纳法""倒推归纳法"，讲"不等式"的时候发了很厚的讲义，均取材于 Hardy、Littlewood 和 Pólya 合著的《不等式》的部分内容.

关先生讲课精益求精，解析透彻，富于启发，板书工整，说话流利，言简意赅. 我当他的助教长达四年，他为同学讲课我每次必到，这是我的任务和工作. 从他那里我学到很宝贵的知识，从数学思维到教学方法，使我终身受益.

依我的体会，"关龙"有以下的长处：

1. 理论严谨

关先生所著的《拓扑空间引论》(科学出版社，1958 年) 在那个时代是最新的论著；他所写的《泛函分析讲义》(高等教育出版社，1958 年) 在那个时候是一本最详细、最完备的标准著作. 1959 级学生刚上大学,就学习严格的实数理论,用商空间和根子空间证明 Jordan 标准型,学习 Lebesgue 积分的新处理,等等,既严格又新颖. 我印象最深的是,波兰数学家 Micusinski 的《算子演算》有俄文与英文译本,关先生建议我读这本书. 关先生说:"从抽象代数域的扩张,联系到算子演算,很出人意料,他的算子演算比 Laplace 变换更为适用." 这后来写到 1959 级的讲义中. 他最擅长的"泛函分析"在他的《高等数学教程》和讲义中表现得得心应手.

2. 注重联系物理

关先生非常重视数学和物理的联系. 他经常说,纯粹数学要搞,但搞的人不要太多. 他经常强调,数学与物理的结合,能促进国防建设、国计民生、能源开发,国家必须富强起来,不然的话只能受气挨打. 他再三提醒辅导教师,要多出一些应用题给学生,用数学来解释物理现象.

他特别重视微分方程和偏微分方程的教学,认为它们同物理的联系非常密切. 他很关心"特殊函数"的教学,因为数学物理方程中用到不少特殊函数,相应地,他十分推崇 Whitaker 和 Watson 的《数学分析》一书,认为这本书可以说是"特殊函数大全".

3. 创新精神

数学系这三条"龙"本身就是创新. 没有登高望远的视野,没有总览全局的气魄,没有对数学前沿的洞察力,任何"龙"都是不能腾飞的.

关先生思维敏捷,语言精练,在课堂中从来不讲题外话;对待学生心平气和、循循善诱. 但是,他绝对不是一个"好好先生",说起话来很有原则. 我第一次见到关先生,马上想到南开大学的数学家吴大任先生. 他们两个神态举止有一些相似,说起话来轻声轻气. 我后来问关先生,他笑着回答:"我与吴先生是广东同乡嘛!"

1959 级数学系的学生,都是高考中的佼佼者,才智过人,关先生非常喜欢和欣赏他们. 除了上课之外,他与学生接触不多. 我们助教向他汇报学生的情况,比如习题多少、难易程度等等,他听得特别仔细. 他告诉我们,学生有好的解法,要记下来,要教学相长. 后来,在教室的墙壁上,有什么好的解法和想法,都可以张贴. 课间休息时,关先生浏览学生的墙报,露出满意的微笑.

关先生生活艰苦朴素,我记得他用过的信封,从反面裱糊后再用一次. 他的衣着大方简朴. 令人惊讶的是,在课堂上,他有时候穿着西装、打着领带为学生们讲课. 在今天的中国, 自是司空见惯, 而在 20 世纪 60 年代是绝无仅有的.

我和关先生相处四年, 经常见面, 保持着师生兼同事的友谊, 可谓 "君子之交淡如水". 回想起来, 我们两人从来没有同桌吃过饭. 每逢春节, 我和几个学生到中关村他的家里拜年, 吃两三粒糖也就告辞了. 我参加了关先生和刘翠娥同志的婚礼, 那是 1958 年冬天的事情, 婚礼在数学研究所的大厅里举行, 来宾很多, 也不送礼, 糖块招待. 我记得很有意思的事情是, 冯康先生的婚礼在同一个晚上举行, 一个在二楼, 一个在三楼.

1962 年, 第二条 "关龙" 出世了, 这就是数学系 1962 级的学生. 那时, 1959 级学生的基础课还没有结束. 在关先生的安排下, 林群先生 (后来被选为中国科学院院士) 为 1962 级学生开设 "数学分析", 我开设 "几何与代数", 历时两年. 1964 年 10 月开始, 我参加了两年的 "四清" 工作队, 直到 1966 年 "文化大革命" 开始, 1970 年中国科大下迁合肥, 直到 1976 年粉碎 "四人帮", 我过了 12 年与数学无关的日子, 所谓的 "关龙" 也就渐渐淡出了我的个人历史.

自从 1970 年中国科大离开北京, 我再也没有见过关先生.

1959 级数学系的学生, 一定非常感激关肇直老师. 三年半的基础课, 接着是一年半的专业课, 都是他独立地支撑着数学的大厦. 在这些学生中, 有后来的中国科学院院士、众多的教授和研究员、科学家和工程技术人员. 他真可谓殚思竭虑, 呕心沥血. 从 1958 年到 1964 年, 我度过了我的青年时代, 我同关肇直教授的名字是联系在一起的. 在我的一生中, 我同中国科学技术大学也是紧紧地联系在一起的.

2

大师风范[1]

  1958年春,党中央决定创建中国科学技术大学,随后,全国各大报纸上相继刊登了介绍这所大学各个系科的资料. 兼任数学系主任的是大名赫赫的华罗庚教授. 对于一个学习数学、即将走向工作岗位的青年学生,这一消息真是令人神往. 当时,说实在的,一提起中国科大,我首先就联想起"华罗庚"这三个大字.

  1958年10月,我从南开大学被分配到中国科大,我的神往竟成了现实,真是喜出望外. 到了中国科大之后,在教室里和校园里,见到华先生的身影是经常的事. 那时他还不到50岁,风采照人,一派大学者的风度. 当时我做关肇直先生的助教,华先生上数学系学生的"高等数学",只要课不冲突,我就去旁听华先生的课. 华先生上课十分生动,不拘形式. 他一边思考,一边推导,有时写了一大黑板,居然会推倒重来,再换一种他认为更好的证法. 唯其如此,我更受启发. 只有大师级的人物,才能有胆识、有能力这样做.

  我这里只讲一个关于"综合讨论班"的故事,因为我印象特别深刻. 1964年,在华先生的倡导和主持下,成立了"综合讨论班",每周举行一次,事先发出通知寄给有关单位. "综合"二字体现了不限学科,只要是有兴趣、有意义的数学问题都可以拿来讨论. 我记得,参加讨论班的,除了中国科大的

---

[1] 节选自《中国科学技术大学数学五十年》,第135~136页.

师生之外，还有从中国科学院数学研究所、北京大学等高校来的专家，济济一堂. 安徽大学的青年教师李世雄，还远道从合肥前往参加. 只有华先生这样的数学大师，才能有这么大的凝聚力和号召力.

在我的一本经历过 37 年风霜、已经褪了色的笔记本里，记载着这样一个故事：

在 1964 年 7 月 23 日的"综合讨论班"上，讨论"拉夫伦捷夫方程"，用到以下结果：

设 b_n 为实数，且
$$\sum_{n=0}^{\infty} b_n(\cos n\theta + \sin n\theta) = 0, \quad 0 \leqslant \theta \leqslant \pi.$$

求证：$b_n = 0 (n = 0, 1, 2, \cdots)$.

事隔一周，在 7 月 30 日的讨论班上，北京大学闵嗣鹤先生给出了一个证明. 受他的启发，我突然想起另一个证明，举手之后便走上黑板，结果在推演中算不下去. 我满脸通红，十分尴尬，第一是当众出丑，第二是生怕华先生训斥. 但我的第二种担心是多余的，华先生没有骂我，我走下黑板，他继续平静地点评这个题目.

我很是难受，中午饭都不想吃. 受教务处处长王榆的派遣，我当天夜里要到安徽招生. 中午，我极力来修补我的证明，最后终于成功了. 我写了一封信给华先生，装上了我的证明，从华先生办公室的门缝里塞了进去. 晚上，我就愉快地登上了南下的火车.

等我从外地回来，我看到了华先生的回信. 这是他当天 (7 月 30 日) 回复我的，又提出了一个比我的办法更简洁的证明，只有四行. 在信的最后，他写道："…… 可见开始的想法是迂回曲折的，事后类多'先见之明'，能不怕曲折搞出东西来，再求直通，研究之道在焉."

华老对我的细心呵护、谆谆教导使我终生难忘. 我谨以此回忆来纪念一代宗师、数学泰斗、我们崇敬和景仰的老师华罗庚教授.

3

奇才怪杰　良师益友[①]

——忆曾肯成先生

从1954年到1958年，我在南开大学数学系读书。1957年春天，我选了"半单纯李氏代数的结构"的课程，教师是大名鼎鼎的几何学专家严志达教授. 采用的教材是苏联数学家邓肯的著作，在当时属于尖端性的工作. 对于一个三年级的学生，我实在无法理解，如坠云里雾里. 邓肯的书有中译本(科学出版社)，译者是曾肯成，其他的内容细节统统忘记了.

1957年夏天，"反'右派'斗争"席卷全国，高等学校首当其冲，严志达教授庆幸躲过了一劫. 只是记得严志达说："曾肯成在清华当学生的时候，一到暑假，他想念的数学书实在太多，不知道念哪一本书好，于是决定抓阄见分晓."即使都是事实，充其量是严志达得意地吹捧有才气的"右派"学生曾肯成，算不上多大的事. 这个人一定聪明绝顶，才气过人，这是我第二次知道曾肯成的名字.

1958年10月14日，周永佩和我从南开大学被分配到中国科学技术大学，我们坐火车从天津来到首都北京. 这是一个秋高气爽、风和日丽的日子，能分配到中国科大，那是最幸运的事情.

[①] 节选自《中国科学技术大学数学五十年》，第149～151页.

我很快地投入到了教学工作中，被分配到了原子核物理和原子核工程系四班当助教，主讲老师是关肇直教授.

1958 年底的一天，数学教研室艾提副主任叫我到他的办公室去一趟. 艾主任对我说："教研室新来了一个人，他叫曾肯成，划成了'右派'. 这个人数学很好，但是，暂时不让他接触学生，让他听听关先生的课再说."当我第三次听到曾肯成的名字时，我们两个人初次见面，一切是那样突然. 曾肯成是湖南湘乡人，我是湖南长沙人，但我们两个人交谈时，他从不讲湖南话. 他 1927 年出生，比我大九岁. 他貌不惊人，个子矮小，爱抽香烟，是一个 chain-smoker. 听关先生的课时，他坐在最后一排，捧着一本俄文版的《量子场论》，一边听一边看着关先生的黑板. 曾肯成终究不是"久困之人". 1959 年，非数学系的数学教材亟待完善和规划，曾肯成确是最恰当的人选. 数学教研室的领导对曾肯成也礼遇有加，从不另眼相看. 我与曾肯成经常接触主要是因为中学生数学竞赛的事情. 1962 年，全国政治形势开始回暖，北京市也恢复了中学生数学竞赛，华罗庚教授又出任竞赛委员会主任. 讨论数学竞赛试题，由华先生主持，闵嗣鹤、王寿仁、越民义、万哲先、王元、龚昇和曾肯成诸先生，都是热心的参加者. 更年轻的教师还有姜伯驹、史济怀和我. 曾肯成出的题目标新立异，不落俗套，以有限的中学知识内容，能导演有声有色的话剧. 那几年北京市数学竞赛试题都被《美国数学月刊》译成英文，足见其水平之优异. 讨论会结束之后，中午在西单西南角的同和居饭庄吃饭，也算是对参加命题者的一种酬劳.

在数学竞赛举行前一个月，每个星期天上午，北京市数学会都租一个场地，向参赛学生、中学数学教师做报告. 每个星期都有一名报告人、一个专题，领衔的报告人是华罗庚、吴文俊两位数学大师，还有其他的数学家. 对于中学生来说，参加数学竞赛和聆听精彩讲座同样重要. 数学大师们同中学生近距离直接对话，他们敏锐的眼光、穿透性的洞察力，甚至大师们的风采，都使年轻学子终生难忘.

曾肯成为北京市的中学生做过多次演讲，其中一个讲稿《复数与几何》我印象最为深刻.

1962 年，北京市数学会筹划编辑"数学小丛书"，曾先生建议由我和伍润生以油印讲义《复数与几何》为基础，加以修改和扩充，出一本小册子. 我们欣喜之余，觉得怀疑的是：凭曾先生的知名度、知识和驾驭语言的能力，况且已经有了讲义，写这样一本小书，对他来讲轻而易举. 我们猜测曾先生的意思是: 写这种小书犹如"小菜一碟". 他向来不看重经济利益, 不屑于为之, 还不如让位于更年轻的人为好.

我和伍润生对曾肯成先生的一番美意将永志不忘. 我们的小册子《复数与几何》于 1964 年在北京出版 (人民教育出版社), 该小册子曾多次重印, 2004 年由科学出版社再次重印.

用复数做几何, 在欧美数学书中早已有之. 但是, 用复数做几何我头一次读到的是曾先生的讲义. 复数可以看成平面向量, 复数可以描述距离、角度和面积, 复数可以非常方便地刻画旋转, 因此, 很多几何问题可以非常方便地用复数来证明. (但是, 也有一些几何问题不能用复数来证明, 或者证明比较笨拙.)

从 1964 年开始, 每每遇到一个几何问题, 我都会问问自己: 有没有复数解法? 经过年复一年的积累, 我终于写成《复数计算与几何证题》(上海教育出版社, 1980 年). 在该书的"前言"中, 我一开始便提及"曾肯成教授的指导".

1996 年, 我的《神奇的复数》在中国台湾地区九章出版社出版. 每当玩复数做几何有所心得的时候, 就想起曾先生对我的启蒙性的帮助.

曾肯成教授是一位杰出的数学家, 数学功力深厚, 无论是"应用数学"还是"抽象数学", 十八般武艺样样皆能. 我曾经问曾先生: "你最拿手的研究是什么?" 他不假思索地说: "有限群." 留学苏联时他专攻计算数学, 晚年从事代数编码与保密通信研究, 为国家做出了重大的贡献. 中国科大建校初期, 曾先生为非数学系数学教材建设做出了不可磨灭的贡献. 他热情地帮助青年教师成长, 不计报酬, 诲人不倦. 他是一位优秀教师, 讲课引人入胜, 语言生动丰富. 我清楚地记得, 他在和我交流数学心得时讲过: "Taylor 公式是一元微分学的顶峰." 玩味至今, 他的见地越想越正确! 曾先生的文学涵养极高, 古典诗词闭目成韵, 精通多门外语, 聪明过人.

2004 年 5 月, 曾肯成教授不幸在北京逝世. 史济怀教授和我专程到北京, 以老同事和老朋友的身份, 在八宝山与曾先生做最后的告别. 告别仪式庄严肃穆, 极尽哀荣, 众多的学生、同事向曾先生致敬.

人们看到他的性格, 既有谦卑温顺, 又有桀骜不驯; 他的经历, 既朴实无华, 又带有传奇色彩; 他的气质, 既有横溢的才华, 又是一个地道的凡人, 这些构成了真正的曾肯成.

4

三次 Hermite 多项式的基底

在计算机辅助几何设计中，美国的 Coons 教授提出了一种设计曲线和曲面的方法，人们称之为 Coons 曲面. 在他的著作中，广泛地使用四个三次多项式函数：

$$F_0(x) = 1 - 3x^2 + 2x^3,$$
$$F_1(x) = 3x^2 - 2x^3,$$
$$G_0(x) = x - 2x^2 + x^3,$$
$$G_1(x) = -x^2 + x^3.$$

实际上这四个函数就是三次 Hermite 多项式的基底. 我们把这些基函数结合起来，推导了三次样条函数的连续性方程. 我们认为，在我们的推导中，无需"凑方"的技巧，平铺直叙，干净利索.

1982 年元旦到 8 月底，我来到 Brown 大学应用数学系当访问学者，接待我的人是著名的应用数学家 Philip J. Davis 教授. 当年他讲授"应用逼近论"时，我也在听课. Davis 称赞我的三次样条的推导，认为是最直接的方法，并写入了他的讲义中.

下面的文章原载于：《数学的实践和认识》，1979 年，2 卷第 55～64 页.

关于三次样条函数的两点注记

中国科学技术大学数学系　　常庚哲

1 引　言

关于样条函数的理论和应用，近年来，在国内一些数学刊物上已有详细介绍[1,2]. 本文主要做了两件事：

1. 采用 Hermite 插值基函数推出三次样条的两种节点关系式；

2. 讨论了端点条件对于样条函数的影响，特别地，改进了文献 [3] 的结果.

现在重点说说第一件事. 设有 $n+1$ 个数据点 $(x_i, y_i), i = 0, 1, \cdots, n$，把它画成如下表格：

x_0	x_1	x_2	$\cdots$	x_n
y_0	y_1	y_2	$\cdots$	y_n

(1′)

并称之为插值条件，今要构造一个适合上述条件的 n 次多项式. 众所周知，Lagrange 插值多项式提供了这个问题的一个解答. 这就是说，首先着手解决比上述更具体的插值问题：

x_0	$\cdots$	x_{i-1}	x_i	x_{i+1}	$\cdots$	x_n
0	$\cdots$	0	1	0	$\cdots$	0

并把它的解记为 $w_i(x)$. 显然，多项式 $w_i(x)$ 应有以下形式：

$$w_i(x) = A_i(x - x_0) \cdots (x - x_{i-1})(x - x_{i+1}) \cdots (x - x_n),$$

其中 A_i 为一常数，可由条件 $w_i(x_i) = 1$ 来确定. 所以

$$w_i(x) = \frac{(x - x_0) \cdots (x - x_{i-1})(x - x_{i+1}) \cdots (x - x_n)}{(x_i - x_0) \cdots (x_i - x_{i-1})(x_i - x_{i+1}) \cdots (x_i - x_n)}.$$

这样一来，上述插值问题的解答便是

$$L(x) = y_0 w_0(x) + y_1 w_1(x) + \cdots + y_n w_n(x),$$

这就是著名的 Lagrange 插值多项式,而

$$w_0(x), w_1(x), \cdots, w_n(x)$$

称为 Lagrange 插值基函数.

在参考文献 [4] 中,华罗庚教授把这种解决问题的技巧追溯到我国古代《孙子算经》上的"物不知其数"问题,他指出:这类问题与解法是世界数学史上著名的东西.

"基函数"方法在高等数学的许多分支里都有不同表现. 基函数犹如是"积木块",用它们可以很容易地拼成各式各样的"建筑物". 采用基函数方法来讨论三次样条,思路简单清晰,几乎只留下了直截了当的计算,并且还可以使得两种节点关系式的推导方法一致起来. 笔者认为,这样做对理论推导、实际计算和教学工作,都有一些好处.

2 m-关系式

样条函数是适合插值条件 (1′)、在区间 $[a,b]$(其中 $a=x_0, b=x_n$) 上具有连续二阶导数的分段三次多项式函数. 用 $S(x)$ 来记一个样条函数,并置

$$m_i = S'(x_i), \quad i = 0, 1, \cdots, n,$$

由于 $S(x)$ 在每一个区间 $[x_{i-1}, x_i](i=1,2,\cdots,n)$ 上都是三次多项式,因此可由带一阶导数的插值条件完全确定.

	x_{i-1}	x_i
(函数值)	y_{i-1}	y_i
(一阶导数值)	m_{i-1}	m_i

(2′)

所谓采用基函数方法,就是同时考察四个与 (2′) 有相同类型,但更加具体的插值问题:

(自变量值)	0	1	0	1	0	1	0	1
(函数值)	1	0	0	1	0	0	0	0
(一阶导数值)	0	0	0	0	1	0	0	1

(3′)

4 三次 Hermite 多项式的基底

并用 $F_0(x), F_1(x), G_0(x), G_1(x)$ 分别记这四个问题的解答. 它们都是三次多项式，故有常数元素的四阶方阵 P 使得

$$[\ F_0(x),\ F_1(x),\ G_0(x),\ G_1(x)\ ] = [\ 1,\ x,\ x^2,\ x^3\ ]P, \tag{1}$$

分别用 $x = 0$ 和 $x = 1$ 代入式 (1)，并对照表 (3′) 可得

$$[\ 1\ 0\ 0\ 0\ ] = [\ 1\ 0\ 0\ 0\ ]P, \tag{2}$$

$$[\ 0\ 1\ 0\ 0\ ] = [\ 1\ 1\ 1\ 1\ ]P. \tag{3}$$

在等式 (1) 两方对 x 求导，得

$$[\ F_0'(x),\ F_1'(x),\ G_0'(x),\ G_1'(x)\ ] = [\ 0,\ 1,\ 2x,\ 3x^2\ ]P. \tag{4}$$

又分别用 $x = 0$ 和 $x = 1$ 代入式 (4)，并对照表 (3′) 得

$$[\ 0\ 0\ 1\ 0\ ] = [\ 0\ 1\ 0\ 0\ ]P, \tag{5}$$

$$[\ 0\ 0\ 0\ 1\ ] = [\ 0\ 1\ 2\ 3\ ]P. \tag{6}$$

将式 (2)、(3) 和式 (5)、(6) 合并写为

$$\begin{bmatrix} 1 & 0 & 0 & 0 \\ 1 & 1 & 1 & 1 \\ 0 & 1 & 0 & 0 \\ 0 & 1 & 2 & 3 \end{bmatrix} P = I_4,$$

这里 I_4 表示四阶单位方阵. 于是

$$P = \begin{bmatrix} 1 & 0 & 0 & 0 \\ 1 & 1 & 1 & 1 \\ 0 & 1 & 0 & 0 \\ 0 & 1 & 2 & 3 \end{bmatrix}^{-1} = \begin{bmatrix} 1 & 0 & 0 & 0 \\ 0 & 0 & 1 & 0 \\ -3 & 3 & -2 & -1 \\ 2 & -2 & 1 & 1 \end{bmatrix},$$

代入式 (1) 算出

$$\begin{cases} F_0(x) = 2x^3 - 3x^2 + 1, \\ F_1(x) = -2x^3 + 3x^2, \\ G_0(x) = x(1-x)^2, \\ G_1(x) = x^2(x-1). \end{cases} \tag{7}$$

注意，由式 (7) 可知

$$F_0(x) + F_1(x) \equiv 1. \tag{8}$$

式 (7) 中的四个函数，称为 Hermite 插值基函数，也就是 Coons 在文献 [5] 中使用的所谓"混合函数"(blending functions). 这样一来，插值问题 (2′) 的解答显然是

$$S(x) = y_{i-1} F_0\left(\frac{x - x_{i-1}}{h_i}\right) + y_i F_1\left(\frac{x - x_{i-1}}{h_i}\right) \\ + h_i \left[m_{i-1} G_0\left(\frac{x - x_{i-1}}{h_i}\right) + m_i G_1\left(\frac{x - x_{i-1}}{h_i}\right) \right], \tag{9}$$

其中，$h_i = x_i - x_{i-1}$, $i = 1, 2, \cdots, n$.

不论 $m_0, m_1, \cdots, m_n$ 取什么数值，由式 (9) 分段定义的函数 $S(x)$ 及其一阶导数 $S'(x)$ 在节点 $x_1, x_2, \cdots, x_{n-1}$ 处总是连续的，因此 $S(x)$ 在 $[a,b]$ 上总是一阶连续可导的. 但是，如果随意选取 $m_0, m_1, \cdots, m_n$，则不能保证在这些点上的二阶导数存在. 由式 (9) 算出

$$S'(x) = \frac{1}{h_i}\left[y_{i-1} F_0'\left(\frac{x - x_{i-1}}{h_i}\right) + y_i F_1'\left(\frac{x - x_{i-1}}{h_i}\right) \right] \\ + m_{i-1} G_0'\left(\frac{x - x_{i-1}}{h_i}\right) + m_i G_1'\left(\frac{x - x_{i-1}}{h_i}\right), \tag{10}$$

$$S''(x) = \frac{1}{h_i^2}\left[y_{i-1} F_0''\left(\frac{x - x_{i-1}}{h_i}\right) + y_i F_1''\left(\frac{x - x_{i-1}}{h_i}\right) \right] \\ + \frac{1}{h_i}\left[m_{i-1} G_0''\left(\frac{x - x_{i-1}}{h_i}\right) + m_i G_1''\left(\frac{x - x_{i-1}}{h_i}\right) \right]. \tag{11}$$

由式 (11) 算出

$$S''(x_i - 0) = \frac{1}{h_i^2}[y_{i-1} F_0''(1) + y_i F_1''(1)] \\ + \frac{1}{h_i}[m_{i-1} G_0''(1) + m_i G_1''(1)], \tag{12}$$

由在 (x_i, x_{i+1}) 上与式 (11) 相当的那样一个公式算出

$$S''(x_i + 0) = \frac{1}{h_{i+1}^2}[y_i F_0''(0) + y_{i+1} F_1''(0)] \\ + \frac{1}{h_{i+1}}[m_i G_0''(0) + m_{i+1} G_1''(0)]. \tag{13}$$

由于

$$\begin{cases} F_0''(x) = 12x - 6, \\ F_1''(x) = -12x + 6, \\ G_0''(x) = 6x - 4, \\ G_1''(x) = 6x - 2. \end{cases} \tag{14}$$

命式 (12) 与式 (13) 的右边相等, 并化简得

$$\frac{h_{i+1}}{h_i + h_{i+1}} m_{i-1} + 2m_i + \frac{h_i}{h_i + h_{i+1}} m_{i+1}$$
$$= 3 \left(\frac{h_{i+1}}{h_i + h_{i+1}} \cdot \frac{y_i - y_{i-1}}{h_i} + \frac{h_i}{h_i + h_{i+1}} \cdot \frac{y_{i+1} - y_i}{h_{i+1}} \right),$$

令

$$\begin{cases} \lambda_i = \dfrac{h_{i+1}}{h_i + h_{i+1}}, \quad \mu_i = 1 - \lambda_i, \\ c_i = 3 \left(\lambda_i \dfrac{y_i - y_{i-1}}{h_i} + \mu_i \dfrac{y_{i+1} - y_i}{h_{i+1}} \right), \end{cases} \tag{15}$$

则有

$$\lambda_i m_{i-1} + 2m_i + \mu_i m_{i+1} = c_i, \quad i = 1, 2, \cdots, n-1. \tag{16}$$

式 (16) 中只包含 $n-1$ 个方程, 还不足以确定 $n+1$ 个未知量 $m_0, m_1, \cdots, m_n$, 所以还要加两个端点条件:

(1) 给定 m_0 和 m_n, 这相当于指定样条曲线在首、末两端点的斜率. 这时式 (16) 中已只含 $n-1$ 个未知数, 可以唯一地将它们解出;

(2) 给定端点处的二阶导数

$$S''(x_0) = M_0, \quad S''(x_n) = M_n.$$

故可在式 (13) 中令 $i = 0$, 算出

$$\frac{6}{h_1^2}(y_1 - y_0) - \frac{2}{h_1}(2m_0 + m_1) = M_0,$$

在式 (12) 中令 $i = n$, 算出

$$\frac{-6}{h_n^2}(y_n - y_{n-1}) + \frac{2}{h_n}(m_{n-1} + 2m_n) = M_n,$$

也就是

$$\begin{cases} 2m_0 + m_1 = \dfrac{3}{h_1}(y_1 - y_0) - \dfrac{M_0 h_1}{2}, \\ m_{n-1} + 2m_n = \dfrac{3}{h_n}(y_n - y_{n-1}) + \dfrac{M_n h_n}{2}. \end{cases} \qquad (17)$$

将式 (16) 和式 (17) 联立起来，也能唯一确定全部 m_i，代回式 (9) 便得到样条函数 $S(x)$ 的分段表达式.

3 M-关系式

按照参考文献 [1] 的说法，方程组 (16) 称为 m-关系式，而令 $M_i = S''(x_i), i = 0, 1, 2, \cdots, n$，它们所应满足的方程组则称为 M-关系式.

推导 M-关系式的通常的办法是：利用 $S''(x)$ 在 $[x_{i-1}, x_i]$ 上为线性函数及 $S''(x_{i-1}) = M_{i-1}$ 与 $S''(x_i) = M_i$，可作出 $S''(x)$ 在此区间上的表达式. 然后连续积分两次，确定积分常数之后，得到 $S(x)$ 的表达式再进行讨论. 这种办法与得出 m-关系式的通常的办法是很不相同的.

但是，我们在这里采用的方法则是与前节完全平行的.

考虑带二阶导数的插值问题：

	x_{i-1}	x_i
(函数值)	v_{i-1}	v_i
(二阶导数值)	M_{i-1}	M_i

$(4')$

首先讨论四个与 $(4')$ 有相同类型，但更加具体的插值问题：

(自变量值)	0	1	0	1	0	1	0	1
(函数值)	1	0	0	1	0	0	0	0
(二阶导数值)	0	0	0	0	1	0	0	1

$(5')$

并用 $\overline{F}_0(x), \overline{F}_1(x), \overline{G}_0(x), \overline{G}_1(x)$ 分别记这四个问题的解答. 它们都是三次多项式，故有常系数四阶方阵 Q 使得

$$\begin{bmatrix} \overline{F}_0(x) & \overline{F}_1(x) & \overline{G}_0(x) & \overline{G}_1(x) \end{bmatrix} = \begin{bmatrix} 1 & x & x^2 & x^3 \end{bmatrix} Q, \qquad (18)$$

4 三次 Hermite 多项式的基底

分别用 $x=0$ 和 $x=1$ 代入式 (18)，并对照表 (5′) 可得

$$[\ 1\ \ 0\ \ 0\ \ 0\ ] = [\ 1\ \ 0\ \ 0\ \ 0\ ]Q, \qquad (19)$$

$$[\ 0\ \ 1\ \ 0\ \ 0\ ] = [\ 1\ \ 1\ \ 1\ \ 1\ ]Q. \qquad (20)$$

在等式 (18) 两边对 x 求导两次，得

$$[\ \overline{F}_0''(x),\ \overline{F}_1''(x),\ \overline{G}_0''(x),\ \overline{G}_1''(x)\ ] = [\ 0,\ 0,\ 2,\ 6x\ ], \qquad (21)$$

又分别用 $x=0$ 和 $x=1$ 代入式 (21)，并对照表 (5′) 得

$$[\ 0\ \ 0\ \ 1\ \ 0\ ] = [\ 0\ \ 0\ \ 2\ \ 0\ ]Q, \qquad (22)$$

$$[\ 0\ \ 0\ \ 0\ \ 1\ ] = [\ 0\ \ 0\ \ 2\ \ 6\ ]Q. \qquad (23)$$

将式 (19)、(20) 和式 (22)、(23) 合并写为

$$\begin{bmatrix} 1 & 0 & 0 & 0 \\ 1 & 1 & 1 & 1 \\ 0 & 0 & 2 & 0 \\ 0 & 0 & 2 & 6 \end{bmatrix} Q = I_4,$$

由此可知

$$Q = \begin{bmatrix} 1 & 0 & 0 & 0 \\ 1 & 1 & 1 & 1 \\ 0 & 0 & 2 & 0 \\ 0 & 0 & 2 & 6 \end{bmatrix}^{-1} = \begin{bmatrix} 1 & 0 & 0 & 0 \\ -1 & 1 & -1/3 & -1/6 \\ 0 & 0 & 1/2 & 0 \\ 0 & 0 & -1/6 & 1/6 \end{bmatrix},$$

代入式 (18) 算出

$$\begin{cases} \overline{F}_0(x) = 1-x, \\ \overline{F}_1(x) = x, \\ \overline{G}_0(x) = -\dfrac{x}{6}(x-1)(x-2), \\ \overline{G}_1(x) = \dfrac{x}{6}(x-1)(x+1). \end{cases} \qquad (24)$$

这里也有

$$\overline{F}_0(x) + \overline{F}_1(x) \equiv 1. \qquad (25)$$

插值问题 (4′) 的解答显然是

$$S(x) = y_{i-1}\overline{F}_0\left(\frac{x-x_{i-1}}{h_i}\right) + y_i\overline{F}_1\left(\frac{x-x_{i-1}}{h_i}\right) \\ + h_i^2\left[M_{i-1}\overline{G}_0\left(\frac{x-x_{i-1}}{h_i}\right) + M_i\overline{G}_1\left(\frac{x-x_{i-1}}{h_i}\right)\right], \tag{26}$$

在等式 (26) 两边对 x 求导

$$S'(x) = \frac{1}{h_i}\left[y_{i-1}\overline{F}'_0\left(\frac{x-x_{i-1}}{h_i}\right) + y_i\overline{F}'_1\left(\frac{x-x_{i-1}}{h_i}\right)\right] \\ + h_i\left[M_{i-1}\overline{G}'_0\left(\frac{x-x_{i-1}}{h_i}\right) + M_i\overline{G}'_1\left(\frac{x-x_{i-1}}{h_i}\right)\right]. \tag{27}$$

为了让 $S(x)$ 在节点 $x_1, x_2, \cdots, x_{n-1}$ 处有一阶导数，先由式 (27) 算出

$$S'(x_i-0) = \frac{1}{h_i}\left[y_{i-1}\overline{F}'_0(1) + y_i\overline{F}'_1(1)\right] \\ + h_i\left[M_{i-1}\overline{G}'_0(1) + M_i\overline{G}'_1(1)\right], \tag{28}$$

再算出

$$S'(x_i+0) = \frac{1}{h_{i+1}}\left[y_i\overline{F}'_0(0) + y_{i+1}\overline{F}'_1(0)\right] \\ + h_{i+1}\left[M_i\overline{G}'_0(0) + M_{i+1}\overline{G}'_1(0)\right], \tag{29}$$

并令 $S'(x_i-0) = S'(x_i+0)$. 由于

$$\begin{cases} \overline{F}'_0(x) = -1, \\ \overline{F}'_1(x) = 1, \\ \overline{G}'_0(x) = -\dfrac{3x^2-6x+2}{6}, \\ \overline{G}'_1(x) = \dfrac{3x^2-1}{6}, \end{cases} \tag{30}$$

所以 $S'(x_i-0) = S'(x_i+0)$ 就是

$$\frac{h_i}{6}(M_{i-1}+M_i) + \frac{h_{i+1}}{6}(M_{i+1}+2M_i) = \frac{y_{i+1}-y_i}{h_{i+1}} - \frac{y_i-y_{i-1}}{h_i},$$

经化简后便是

$$\mu_i M_{i-1} + 2M_i + \lambda_i M_{i+1} = d_i, \tag{31}$$

其中

$$d_i = \frac{6}{h_i + h_{i+1}} \left(\frac{y_{i+1} - y_i}{h_{i+1}} - \frac{y_i - y_{i-1}}{h_i} \right), \quad i = 1, 2, \cdots, n-1.$$

如果给定 M_0 和 M_n，则由方程组 (31) 可以唯一解出 $M_1, M_2, \cdots, M_{n-1}$；如果给定两端点处的一阶导数 m_0 和 m_n，则可导出两个方程式

$$2M_0 + M_1 = \frac{6}{h_1} \left(\frac{y_1 - y_0}{h_1} - m_0 \right), \tag{32}$$

$$M_{n-1} + 2M_n = \frac{6}{h_n} \left(m_n - \frac{y_n - y_{n-1}}{h_n} \right). \tag{33}$$

它们与 M-关系式 (31) 合并在一起，可以唯一地解出全部 M_i，再代回公式 (26)，便得到样条函数的分段表达式，这样得出的 $S(x)$，在区间 $[a,b]$ 上是二阶连续可导的.

4 端点条件对样条函数的影响

设想我们要以插值条件 (1′) 与二阶导数端点条件 y_0'' 与 y_n'' 构造样条函数 $S_1(x)$，由于误差，使端点条件分别变为 $y_0'' + \varepsilon_0$ 和 $y_n'' + \varepsilon_n$，而得到的样条函数为 $S_2(x)$，故需要研究函数

$$S(x) = S_2(x) - S_1(x). \tag{34}$$

这个函数适合条件

$$\begin{cases} S(x_0) = S(x_1) = \cdots = S(x_n) = 0, \\ S''(x_0) = \varepsilon_0, \quad S''(x_n) = \varepsilon_n. \end{cases} \tag{35}$$

我们另外定义两个样条函数 $\psi_0(x)$ 和 $\psi_n(x)$，它们满足

$$\begin{cases} \psi_0(x_i) = \psi_n(x_i) = 0, \quad 0 \leqslant i \leqslant n, \\ \psi_0''(x_0) = 1, \quad \psi_0''(x_n) = 0, \\ \psi_n''(x_0) = 0, \quad \psi_n''(x_n) = 1, \end{cases}$$

于是显然有

$$S(x) = \varepsilon_0 \psi_0(x) + \varepsilon_n \psi_n(x). \tag{36}$$

先考察 $\psi_0(x)$，令 $k_i = \psi_0'(x_i), 0 \leqslant i \leqslant n$，由 m-关系式及式 (17) 知 $k_0, k_1, \cdots, k_n$ 应适合方程组

$$\begin{bmatrix} 2 & 1 & & & & \\ \lambda_1 & 2 & \mu_1 & & & \\ & \ddots & \ddots & \ddots & & \\ & & \ddots & \ddots & \ddots & \\ & & & \ddots & \ddots & \ddots \\ & & & \lambda_{n-1} & 2 & \mu_{n-1} \\ & & & & 1 & 2 \end{bmatrix} \begin{bmatrix} k_0 \\ k_1 \\ \vdots \\ \vdots \\ \vdots \\ k_{n-1} \\ k_n \end{bmatrix} = \begin{bmatrix} \dfrac{-h_1}{2} \\ 0 \\ \vdots \\ \vdots \\ \vdots \\ 0 \\ 0 \end{bmatrix}. \tag{37}$$

从最后一个方程解得

$$k_n = -p_{n-1} k_{n-1},$$

其中 $p_{n-1} = \dfrac{1}{2}$.

将此代入前头一个方程，解得

$$k_{n-1} = -p_{n-2} k_{n-2},$$

其中

$$p_{n-2} = \dfrac{\lambda_{n-1}}{2 - p_{n-1} \mu_{n-1}},$$

继续往前代入，一般有

$$k_i = -p_{i-1} k_{i-1}, \quad i = 1, 2, \cdots, n, \tag{38}$$

其中

$$\begin{cases} p_{i-1} = \dfrac{\lambda_i}{2 - p_i \mu_i}, & 1 \leqslant i \leqslant n-1, \\ p_{n-1} = \dfrac{1}{2}. \end{cases} \tag{39}$$

另外，由式 (38) 得出

$$k_i = (-1)^i p_0 p_1 \cdots p_{i-1} k_0, \quad 1 \leqslant i \leqslant n. \tag{40}$$

现在来证明

$$0 < p_{i-1} \leqslant \dfrac{1}{2}, \quad 1 \leqslant i \leqslant n.$$

这是因为首先有 $p_{n-1} = \dfrac{1}{2}$，而当 $0 < p_i \leqslant \dfrac{1}{2}$ 时，

$$0 < \frac{\lambda_i}{2} < p_{i-1} = \frac{\lambda_i}{2 - p_i \mu_i} \leqslant \frac{\lambda_i}{2 - 0.5\mu_i} = \frac{2\lambda_i}{4 - \mu_i}$$

$$= \frac{2\lambda_i}{3 + \lambda_i} < \frac{2\lambda_i}{3\lambda_i + \lambda_i} = \frac{1}{2}.$$

这样一来，由式 (40) 可推出

$$|k_i| = p_0 p_1 \cdots p_{i-1}|k_0| \leqslant \left(\frac{1}{2}\right)^i |k_0|. \tag{41}$$

今往估计 $|k_0|$。用 $k_1 = -p_0 k_0$ 代入式 (37) 的第一个方程 $2k_0 + k_1 = -\dfrac{h_1}{2}$，解出

$$k_0 = -\frac{h_1}{2(2 - p_0)} < 0,$$

所以

$$|k_0| = \frac{h_1}{2(2 - p_0)} < \frac{h_1}{2\left(2 - \dfrac{1}{2}\right)} = \frac{h_1}{3},$$

于是，从式 (41) 得到

$$|k_i| < \frac{1}{3} h_1 \left(\frac{1}{2}\right)^i, \quad i = 0, 1, 2, \cdots, n. \tag{42}$$

另一方面，若令 $l_i = \psi'_n(x_i)$，由对称性的考虑，参照式 (42) 可得

$$|l_i| < \frac{1}{3} h_n \left(\frac{1}{2}\right)^{n-i}, \quad i = 0, 1, 2, \cdots, n. \tag{43}$$

对由公式 (36) 确定的函数，有

$$m_i = S'(x_i) = \varepsilon_0 k_i + \varepsilon_n l_i,$$

于是由式 (42) 和式 (43) 得到

$$|m_i| < \frac{\left[|\varepsilon_0| h_1 \left(\dfrac{1}{2}\right)^i + |\varepsilon_n| h_n \left(\dfrac{1}{2}\right)^{n-i}\right]}{3}, \quad i = 0, 1, 2, \cdots, n. \tag{44}$$

再讨论 ε_0 和 ε_n 对 $S(x)$ 的节点上二阶导数 M_i 的影响. 令 $K_i = \psi_0''(x_i)$, 此时 $K_0 = 1$, $K_n = 0$, 所以确定 $K_0, K_1, \cdots, K_{n-1}$ 的方程组为

$$\begin{bmatrix} 2 & 0 & & & & \\ \mu_1 & 2 & \lambda_1 & & & \\ & \ddots & \ddots & \ddots & & \\ & & \ddots & \ddots & \ddots & \\ & & & \ddots & \ddots & \ddots \\ & & & & \mu_{n-2} & 2 & \lambda_{n-2} \\ & & & & & \mu_{n-1} & 2 \end{bmatrix} \begin{bmatrix} K_0 \\ K_1 \\ \vdots \\ \vdots \\ \vdots \\ K_{n-2} \\ K_{n-1} \end{bmatrix} = \begin{bmatrix} 2 \\ 0 \\ \vdots \\ \vdots \\ \vdots \\ 0 \\ 0 \end{bmatrix}. \quad (45)$$

由于此时有

$$K_{n-1} = -\frac{\mu_{n-1}}{2} K_{n-2}, \quad 0 < \frac{\mu_{n-1}}{2} < \frac{1}{2},$$

故仿照前面的推导可得出

$$|K_i| \leqslant |K_0| \left(\frac{1}{2}\right)^i = \left(\frac{1}{2}\right)^i, \quad 0 \leqslant i \leqslant n.$$

再令 $L_i = \psi_n''(x_i)$, 由对称性的考虑可知

$$|L_i| \leqslant \left(\frac{1}{2}\right)^{n-i}, \quad 0 \leqslant i \leqslant n.$$

由于 $M_i = S''(x_i) = \varepsilon_0 K_i + \varepsilon_n L_i$, 故

$$|M_i| < |\varepsilon_0| \left(\frac{1}{2}\right)^i + |\varepsilon_n| \left(\frac{1}{2}\right)^{n-i}, \quad i = 0, 1, 2, \cdots, n. \quad (46)$$

我们把式 (44) 和式 (46) 合起来写成:

定理 1 设 $S(x)$ 是适合 $S(x_i) = 0$, $0 \leqslant i \leqslant n$ 及 $S''(x_0) = \varepsilon_0$, $S''(x_n) = \varepsilon_n$ 的三次样条函数, 那么 $S(x)$ 在各节点上的一阶导数和二阶导数将满足不等式

$$|m_i| < \frac{\left[|\varepsilon_0|h_1\left(\frac{1}{2}\right)^i + |\varepsilon_n|h_n\left(\frac{1}{2}\right)^{n-i}\right]}{3},$$

$$|M_i| < |\varepsilon_0|\left(\frac{1}{2}\right)^i + |\varepsilon_n|\left(\frac{1}{2}\right)^{n-i}, \quad i = 0, 1, 2, \cdots, n.$$

这显然是比参考文献 [3] 中的定理 1 和定理 2 更强的结果.

如果要讨论一阶导数的端点条件由于误差而产生的对于样条函数的影响，则应研究函数

$$S(x) = \varepsilon_0 \varphi_0(x) + \varepsilon_n \varphi_n(x), \tag{47}$$

其中，$\varphi_0(x)$ 和 $\varphi_n(x)$ 是两个样条函数，适合

$$\begin{cases} \varphi_0(x_i) = \varphi_n(x_i) = 0, & i = 0, 1, \cdots, n, \\ \varphi_0'(x_0) = 1, \quad \varphi_0'(x_n) = 0, \\ \varphi_n'(x_0) = 0, \quad \varphi_n'(x_n) = 1. \end{cases} \tag{48}$$

设 $k_i = \varphi_0'(x_i)$，其中，$k_0 = 1$, $k_n = 0$，用完全类似的做法可以证明

$$|k_i| \leqslant \left(\frac{1}{2}\right)^i, \quad 0 \leqslant i \leqslant n. \tag{49}$$

又令 $l_i = \varphi_n'(x_i)$，其中 $l_0 = 0$, $l_n = 1$，由对称性的考虑可知有

$$|l_i| \leqslant \left(\frac{1}{2}\right)^{n-i}, \quad 0 \leqslant i \leqslant n. \tag{50}$$

再令 $K_i = \varphi_0''(x_i)$，那么由 M-关系式与 (32)、(33) 两式，知应有

$$\begin{bmatrix} 2 & 0 & & & & \\ \mu_1 & 2 & \lambda_1 & & & \\ & \ddots & \ddots & \ddots & & \\ & & \ddots & \ddots & \ddots & \\ & & & \ddots & \ddots & \ddots \\ & & & \mu_{n-1} & 2 & \lambda_{n-1} \\ & & & & 1 & 2 \end{bmatrix} \begin{bmatrix} K_0 \\ K_1 \\ \vdots \\ \vdots \\ \vdots \\ K_{n-1} \\ K_n \end{bmatrix} = \begin{bmatrix} -\dfrac{6}{h_1} \\ 0 \\ \vdots \\ \vdots \\ \vdots \\ 0 \\ 0 \end{bmatrix},$$

此时可以推出

$$|K_i| \leqslant \left(\frac{1}{2}\right)^i |K_0|,$$

而

$$|K_0| < \frac{4}{h_1},$$

所以

$$|K_i| < \frac{4\left(\frac{1}{2}\right)^i}{h_1}, \quad 0 \leqslant i \leqslant n. \tag{51}$$

最后再令 $L_i = \varphi_n''(x_i)$，将得

$$|L_i| < \frac{4\left(\frac{1}{2}\right)^{n-i}}{h_n}, \quad 0 \leqslant i \leqslant n. \tag{52}$$

把式 (49)、(50)、(51)、(52) 综合起来，有：

定理 2　设 $S(x)$ 是适合 $S(x_i) = 0, 0 \leqslant i \leqslant n$ 及 $S'(x_0) = \varepsilon_0, S'(x_n) = \varepsilon_n$ 的三次样条函数，那么 $S(x)$ 在各节点上的一阶导数和二阶导数将满足不等式

$$|m_i| < |\varepsilon_0|\left(\frac{1}{2}\right)^i + |\varepsilon_n|\left(\frac{1}{2}\right)^{n-i},$$

$$|M_i| < 4\left[\frac{|\varepsilon_0|\left(\frac{1}{2}\right)^i}{h_1} + \frac{|\varepsilon_n|\left(\frac{1}{2}\right)^{n-i}}{h_n}\right].$$

参 考 文 献

[1]　孙家昶. 样条函数及其在计算方法上的某些应用 (内部资料), 1974.

[2]　潘承洞. Spline 函数的理论及其应用 (二)[J]. 数学的实践与认识，1975(4)：56 − 77.

[3]　保明堂. 自然三次样条在增压器叶轮二元流设计计算中的误差估计[J]. 数学的实践与认识，1978(1)：41 − 50.

[4]　华罗庚. 我国古代数学成就之一瞥[J]. 文物，1978(1)：46 − 49.

[5]　Coons S A, AD-663504.

5

Pedoe 不等式

著名的几何学家 Daniel Pedoe (1910~1998) 在伦敦出生，后来移居美国. 我在南开大学数学系当学生的时候，就看过 Pedoe 教授的书. 一本叫《The Gentle Arts of Mathematics》，是一本通俗读物，从校图书馆借来的，目的是想学点英文. 另外一本书是 Hodge 与 Pedoe 合著的《代数几何方法》(第 1 卷，共 3 卷) 俄文译本，我们年级的代数讨论班就念这本书. 实际上是线性代数和抽象代数，代数几何还没有入门.

在《美国数学月刊》上，经常见到 Pedoe 教授的论文和提出来的几何题. 其中有一个牵涉两个三角形边长和面积的不等式，这就是人们所称的 Pedoe 不等式，被 Pedoe 教授所钟爱. 他用不同的方法来证明这个不等式，证明巧妙，非常优美，都是纯几何的方法.

我提出的 Pedoe 不等式的复数证明，在《美国数学月刊》1982 年，89 卷第 692 页上发表. 用复数证明几何题，不需要巧思，只需要踏踏实实、直截了当地计算.

我的小小的文章，惊动了数学大师陈省身 (1911~2004)，使我受宠若惊. 1978 年，我国选派第一批人员到美国当访问学者，我校数学系彭家贵教授，幸运地选上了，接待人是加州大学伯克利分校的陈省身教授. 我对他讲，我想在美国发表文章，不得其门而入，请他费心帮助一下，根本没有想到从未谋面陈先生会帮忙. 有一天我收到彭家贵从美国寄来的信，简短说明原委后，下面就是陈省身先生的复印件，从英文翻译成中文，就是：

亲爱的拉尔夫：

　　送上常庚哲的文章《用复数计算证明 Pedoe 不等式》，他想投寄《美国数学月刊》，请你酌情刊登．未来的通信请直接寄到：
　　中华人民共和国　安徽　合肥　中国科学技术大学　数学系　常庚哲教授

顺致最好的祝愿！

陈省身
1980 年元月 28 日

　　我猜想，拉尔夫是《美国数学月刊》的主编．万万没想到，没有任何分量的习作，有劳陈先生的大驾．但是，我终究见过陈先生，而且我还差一点有机会聆听陈先生的教导．

　　1985 年 6 月 12 日晚上，数学大师华罗庚教授在日本东京逝世，他的多年老友、数学大师陈省身和夫人正在从北京到合肥的软卧列车上．列车早上的新闻联播，播出了华罗庚先生去世的消息，陈先生为之悲伤与震惊．陈先生夫妇是中国科大学请来的贵宾．中午在东区图书馆接见教授和学生，和陈先生合影留念．陈先生发言，谈到他与华先生的友谊，无不为之动容．散会后，又被许多教授和学生所簇拥，我来不及说一声道谢．五年前的"推荐信"，陈先生一定不记得了．

　　又一个能够接近陈先生的机会，让我慢慢细说．2004 年，南开大学数学研究所出版了一本数学月历，"总策划"是陈先生，名叫《数学之美》，陈先生题字．一年有 12 个月，每个月都有一个标题，浓缩到 12 页，只有陈先生有这样的胆识和智慧．这 12 个标题是：

　　1. 复数
　　2. 正多面体
　　3. 刘徽与祖冲之
　　4. 圆周率的计算
　　5. 高斯
　　6. 圆锥曲线
　　7. 双螺旋线

8. 国际数学家大会 (在北京举行)
9. 计算机的发展
10. 分形
11. 麦克斯韦方程
12. 中国剩余定理

由于"数学挂历"取得巨大成功，供不应求，陈先生马上想到，写一本书，每月介绍相关的数学内容，每篇有 1 万字左右，由不同的行家执笔，交由科学出版社出版. 齐东旭教授认识陈先生，第一章"复数"由齐执笔. 承蒙东旭的美意，要我们共同编写，我欣然同意. 我渴望见到这位大师，想得到他的指教. 东旭与我很快拿出了初稿，请陈先生斧正，得到他的首肯. 令人痛心的是，2004 年 12 月 3 日陈先生病逝于天津，写小册子的事情化为泡影.

在中国，我在为中学老师和学生的讲演中，宣传 Pedoe 不等式. 后来，我与 Pedoe 教授取得了联系. 他的来信很长，像一个老朋友一样，无话不谈. 他给我的信，我全都珍藏着. 他送给我他的著作，上面有他的签名，穿越太平洋寄往中国. 我们两个未曾谋面，甚是遗憾. 他送给我唯一的一张他的照片在我合肥的家里.

在 1980 年代，杨路教授、张景中教授和我，在中国科大数学系共事. 后来张景中被选为中国科学院院士. 我与 Pedoe 教授相识非常浅，而他们与 Pedoe 是业务上最好的朋友、最好的同行. 我们都知道，Pedoe 教授是杰出的几何学家. 杨与张将 Pedoe 不等式推广到高维. 在他的晚年，也许他最关心的是"生锈圆规"作图. 杨、张的一系列工作，让这位几何学家兴奋不已. 我不妨引用 Pedoe 教授的一段话，在他的《我爱上了几何》[①]论文中，Pedoe 教授写道，"为了让更多的人知晓这个定理，《Crux Mathematicorum》的读者有一些是几何专家. 不久前，我听到常庚哲在 Brown 大学当访问学者，他寄给我这个定理证明中 (b) 部分的英文翻译，作者是中华人民共和国的杨路和张景中. 当 Freeman Dyson 看见他们的证明时，强调说：他们的证明比我的证明聪明很多. 他们的证明发表在《Crux Mathematicorum》，1982 年，8 卷第 79 页上.

......

[①] Pedoe D. In Love with Geometry[J]. The College Mathematics Journal, 1988, 29(3):170-188.

显然，杨和张是在'文化大革命'结束之后才与西方学者取得联系的. 我知道杨路在这历史事件中所经历的苦难，令人难以置信. 但是当秩序重返中国大陆时，他找到了中国科学院的教职. 他曾经访问过美国，成为一位用计算机自动证明几何定理的杰出研究者. 杨与张发表了很多文章，明显表明他们对几何和数学分支的掌握，依我之见，远超过西方数学家的主流."

下面的短文原载于:《The American Mathematical Monthly》，1982 年，89 卷第 692 页.

Proving Pedoe's Inequality by Complex Number Computation

Geng-zhe Chang

Division of Applied Mathematics, Brown University, Providence, RI 02912

Let a, b, c denote the sides of the triangle ABC and let a', b', c' denote the sides of the triangle $A'B'C'$. Let F and F' denote the respective areas of these triangles. Pedoe [1], [2] has proved that

$$a'^2(-a^2+b^2+c^2) + b'^2(a^2-b^2+c^2) + c'^2(a^2+b^2-c^2) \geqslant 16FF' \qquad (1)$$

with equality if and only if the triangles ABC and $A'B'C'$ are similar. Carlitz has verified (1) algebraically in [3] and given a further discussion in [4]. We now present another verification of (1) by complex-number computation.

Set the vertex C of triangle ABC on the origin of the complex plane, and denote the other vertices by complex numbers A and B, the same symbols of these vertices for simplification. Deal with triangle $A'B'C'$ similarly. Hence

$$a = |B|, \quad b = |A|, \quad c = |A-B|, \quad a' = |B'|, \quad b' = |A'|, \quad c' = |A'-B'|.$$

Therefore

$$a'^2(-a^2+b^2+c^2) = B'\overline{B'}[2A\overline{A} - (A\overline{B}+\overline{A}B)],$$
$$b'^2(a^2-b^2+c^2) = A'\overline{A'}[2B\overline{B} - (A\overline{B}+\overline{A}B)],$$
$$c'^2(a^2+b^2-c^2) = [A'\overline{A'}+B'\overline{B'} - (A'\overline{B'}+\overline{A'}B')](A\overline{B}+\overline{A}B).$$

Adding both sides in the above three equalities, we obtain

$$H = 2(|A'|^2|B|^2 + |A|^2|B'|^2) - (A\overline{B}+\overline{A}B)(A'\overline{B'}+\overline{A'}B'),$$

where H denotes the quantity on the left-hand side of (1).

For the area F of triangle ABC, we have

$$F = \frac{1}{2}|I(\overline{A}B)| = \left|\frac{\overline{A}B - A\overline{B}}{4i}\right|$$

and similarly for F'. Then $16FF' = \pm(\overline{A}B - A\overline{B})(\overline{A'}B' - A'\overline{B'})$, the sign being chosen to make the expression positive. Then

$$H - 16FF' = 2|AB' - A'B|^2 \quad \text{or} \quad 2|A\overline{B'} - \overline{A'}B|^2$$

according to whether the triangles ABC and $A'B'C'$ have the same orientation or not. Thus $H \geqslant 16FF'$, with equality if and only if

$$\frac{B}{A} = \frac{B'}{A'} \quad (\text{when the two triangles have same orientation})$$

or

$$\frac{B}{A} = \overline{\left(\frac{B'}{A'}\right)} \quad (\text{when the two triangles have opposite orientation}).$$

These two equalities are equivalent to the condition that the triangles ABC and $A'B'C'$ are similar. This completes the proof of Pedoe's inequality.

References

[1] Pedoe D. Proc. Cambridge Philos[J]. Soc., 1943, 38: 397.
[2] Pedoe D. this Monthly. 1963, 70: 1012.
[3] Carlitz L. this Monthly. 1971, 78: 772.
[4] Carlitz L. this Monthly. 1973, 80: 910.

6

Douglas–Neumann 定理

在初等几何中,有一个定理:在任一个三角形的三边上,各自向外作一个正三角形,那么把这三个三角形的重心连接起来,便得到了一个正三角形.

把命题中的"向外"一律改成"向内",结论照旧是正确的.

上述命题也可以等价地改述为:在任意一个三角形三条边上,以各条边为底向外各作一个底角为 $30°$ 的等腰三角形,那么把这个三角形的新顶点(称为"自由顶点")连接起来,便得到了一个等边三角形.

这就是 Napolean 定理. 可以认为, 这个定理提供了从任一三角形出发, 来构造一个等边三角形的步骤.

Napolean 定理的内容被极大地发展和丰富了. 1940 年 J. Douglas 和 1941 年 B. H. Neumann 分别独立地将上述命题推广到任意多边形的情况, 提出了由任意的 n 边形出发, 来构造一个正 n 边形的一种确定的步骤. Neumann 在 1942 年的证明中, 使用了复数的方法, 使得他的证明非常漂亮, 直观动态.

1981 年夏天, 应 Barnhill 教授的邀请, Brown 大学应用数学系 Davis 教授来到 Utah 大学数学系访问. 久闻他的大名, 这是我第一次见到他. 我对他说, 我想到他们系里访问. 后来, 谈起他的新作《Circulant Matrices》, 有一个习题关于三角形套的收敛估计, 要我精确估算一下. 他回到 Brown 大学之后, 我的结果从邮局已经寄出了. Davis 回信写道: "非常优雅漂亮. 我们可能合写一篇文章?" 我非常兴奋. 转念一想, 这么个小问题, 相当初等, 内容单薄，怎么能够写成文章？全靠 Davis 教授的生花之笔.

以后，我和 Davis 经常通信联系，我把我的关于 Douglas 和 Neumann 定理的预印本给他，请他提意见. 说来也巧，Davis 和 Neumann 都是老朋友，一个在美国，一个在澳大利亚. 那时候，Neumann 教授是 Houston 数学杂志的编委. 我的预印本从 Davis 手中寄到 Neumann 手中，直到 1982 年见报，几乎是最快的速度. 这与 Neumann 教授的推荐有很大关系.

后来，我和 Davis 教授进一步推广这一定理，发表在《Linear Algebra and its Application》, 1983, 54:87~95 上.

我有幸见到 Neumann 教授，是在澳大利亚的悉尼. 1988 年 7 月 9 日晚上，在悉尼的基督教女青年会为欢迎前来参加第 29 届国际数学奥林匹克竞赛的各国领队的宴会上，我见过他，当时他已八十高龄，但是他身体非常硬朗，还经常往来于世界各地.

下面的文章原载于：《Houston Journal of Mathematics》, 1982 年，第 15~18 页.

A Proof of A Theorem of Douglas and Neumann by Circulant Matrices

Geng-zhe Chang [1]

University of Science and Technology of China, Hefei, Anhui, China

Given any triangle, erect on its sides outwardly (or inwardly) equilateral triangles. Then the centers of the three equilateral triangles form an equilateral triangle. This is known as *Napoleon's theorem* (see [1]). The theorem provides a construction of an equilateral triangle from an arbitrarily given triangle.

In [2] and [3], Douglas and Neumann independently extended Napoleon's theorem to the case of polygons. They provided a construction of a regular polygon from an arbitrarily given polygon.

There are many proofs of their statement in [2], [3], [4] and [5]; perhaps the simplest one is given by Neumann in [5]. Using circulant matrices we provide a different aspect of his proof.

Following [1], let the complex numbers $z_1, z_2, \cdots, z_n$ denote the ordered vertices of a polygon P. We denote the polygon by the $n \times 1$ matrix

$$P = \begin{bmatrix} z_1 \\ z_2 \\ \vdots \\ z_n \end{bmatrix}.$$

Let c be a complex number. Now erect on the sides of P triangles that are directly similar to the triangle $01c$. For example, on the side $z_1 z_2$ of P this would be a triangle $z_1 z_2 z_{12}$ where z_{12} is called a free vertex of the triangle. Then z_{12} is given by

$$z_{12} = (1-c)z_1 + c z_2.$$

---

[1] Visiting scholar at the Department of Mathematics, University of Utah, Salt Lake City, Utah 84112.

The n free vertices of n triangles erected on n sides of P form a polygon

$$P_1 = \begin{bmatrix} 1-c & c & & & & \\ 0 & 1-c & c & & & \\ \vdots & & \ddots & \ddots & \ddots & \\ 0 & & & \ddots & \ddots & c \\ c & 0 & \cdots & & 0 & 1-c \end{bmatrix} P.$$

Denote the above circulant matrix of order n by K_c. Then

$$P_1 = K_c P. \tag{1}$$

Rewrite

$$K_c = c(\lambda_c I + \Pi),$$

where I is the identity matrix of order n, $\lambda_c = \dfrac{(1-c)}{c}$ and

$$\Pi = \begin{bmatrix} 0 & 1 & 0 & \cdots & 0 \\ 0 & 0 & 1 & \cdots & 0 \\ \vdots & \vdots & \vdots & & \vdots \\ 0 & 0 & 0 & \cdots & 1 \\ 1 & 0 & 0 & \cdots & 0 \end{bmatrix},$$

(1) thus becomes

$$P_1 = c(\lambda_c I + \Pi)P. \tag{2}$$

Repeating the same process with P_1 we obtain a polygon P_2, and so forth, but the triangle used in each step may be different, that is, the complex numbers c may vary from step to step. Suppose in j-th step we use the complex number c_j. If m processes have been done, using (2) we obtain

$$P_m = c_1 \cdots c_m (\lambda_1 I + \Pi) \cdots (\lambda_m I + \Pi) P, \tag{3}$$

where we use λ_j for λ_{c_j}. Write

$$(\lambda_1 I + \Pi) \cdots (\lambda_m I + \Pi) = d_0 + d_1 \Pi + \cdots + d_m \Pi^m. \tag{4}$$

6 Douglas-Neumann 定理

Now we show that for $m = n-1$ we can choose $\lambda_1, \lambda_2, \cdots, \lambda_{n-1}$ in such a way that $d_0 = d_1 = \cdots = d_{n-1} = 1$.

Set $\lambda_j = -\omega^j \, (j = 1, 2, \ldots, n-1)$, where

$$\omega = \cos\frac{2\pi}{n} + i\sin\frac{2\pi}{n}.$$

Then

$$(t+\lambda_1)\cdots(t+\lambda_{n-1}) = (t-\omega)\cdots(t-\omega^{n-1}) = \frac{t^n-1}{t-1} = 1+t+\cdots+t^{n-1}.$$

Using (3) and (4) we have

$$P_{n-1} = c_1\cdots c_{n-1}(1+\Pi+\cdots+\Pi^{n-1})P$$

$$= c_1\cdots c_{n-1}\begin{bmatrix} 1 & 1 & \cdots & 1 \\ 1 & 1 & \cdots & 1 \\ \vdots & \vdots & & \vdots \\ 1 & 1 & \cdots & 1 \end{bmatrix}P.$$

Since $c_j = 1/(1+\lambda_j) = 1/(1-w^j)\,(j=1,2,\cdots,n-1)$, we have that

$$c_1\cdots c_{n-1} = \frac{1}{(1-\omega)\cdots(1-\omega^{n-1})} = \left(\frac{1}{1+t+\cdots+t^{n-1}}\right)_{t=1} = \frac{1}{n}$$

and

$$P_{n-1} = \begin{bmatrix} z^* \\ z^* \\ \vdots \\ z^* \end{bmatrix},$$

where $z^* = \dfrac{(z_1+z_2+\cdots+z_n)}{n}$ is the center of gravity of the original polygon P.

It is easy to show that

$$|1-c_j| = |c_j|,$$
$$\arg c_j = \frac{\pi}{2} - j\frac{\pi}{n}, \quad j = 1, 2, \cdots, n-1.$$

This means that all the triangles $01c_j$ are isosceles with base angles $\frac{\pi}{2} - j\frac{\pi}{n}$.

Thus, after $n-1$ steps the polygon P_{n-1} reduces to a single point. Omitting any one of the $(n-1)$ steps we obtain a polygon such that, if certain similar triangles are erected on its sides as bases the n free vertices all coincide. The polygons are therefore regular.

This is the theorem of Douglas and Neumann as stated in [3]:

Theorem 1 *If isosceles triangles with base angles $\frac{\pi}{2} - j\frac{\pi}{n}$ are erected on the sides of an arbitrary n-polygon P, and if this process is repeated with the polygon formed by the free vertices of the triangles, but with a different value of j, and so on until any (n−2) values of j from 1,2,$\cdots$,n−1 have been used in arbitrary order, then a regular polygon is obtained.*

Napoleon's theorem is now a special case (for $n = 3$) of the above theorem.

References

[1]   Davis P J. Circulant Matrices[M]. John Wiley & Sons, 1979.

[2]   Douglas J, Geometry of Polygons in the Complex Plane[J]. J. of Math. and Physics, 1940, 19: 93-130.

[3]   Neumann B H. Some Remarks on Polygons[J]. London Math. Soc. J., 1942, 16: 230-245.

[4]   Baker H F. A Remark on Polygons[J]. London Math. Soc. J., 1942, 17: 162-164.

[5]   Neumann B H. A Remark on Polygons[J]. London Math. Soc. J., 1942, 17: 165-166.

[6]   Douglas J. On Linear Polygon Transformations[J]. Bull. Amer. Math. Soc., 1940,46: 551-560.

7

三角形曲面

1974年，在美国盐湖城Utah大学召开的国际会议，由数学系R. E. Barnhill教授和计算机系R. F. Riesenfeld教授共同主持，出版了会议论文集. 这个会议之所以重要，是因为会议明确宣告"计算机辅助几何设计"（Computer Aided Geometric Design，简称CAGD）以学科名义正式诞生了. 论文贡献者中，还有CAD巨匠——美国的S. A. Coons和法国的P. Bézier. 有很多的数学家也参加了这次大会.

这本会议文集慢慢地走到中国，首先感兴趣的人士有航空界、造船界和机械制造的工程师和专家. 北京的《国外航空》编辑部牵头翻译成中文，有三所航空院校、中国科大和中国科学院，参加的人士众多. 这本书没有公开发售，是一种内部资料.

可以肯定，有不少的人由于这一本论文集，对CAGD产生特别的兴趣，以至于作为终身职业的，大有人在. 浙江大学数学系的梁友栋先生和我本人，就是两个例子. 我同梁先生第一次见面是在杭州西湖刘庄，1979年初，他主持浙大一个会议. 那年秋天，他到美国盐湖城Utah大学当访问学者，主人是Riesenfeld.

1980年8月12日，我离开北京，目的地是美国盐湖城. 那时候中国和美国没有开通直航，在法国巴黎转飞机，在旅店住一夜，第二天飞往华盛顿，在中国大使馆住宿. 我们中国科学院留学人员浩浩荡荡，有30多人，我们的同事杨劲根和我都是团员，他到美国是去读研究生的.

　　从华盛顿到盐湖城，还要在芝加哥转飞机. Barnhill 教授在盐湖城机场接我，他是 1939 年生的，与我一样，但头发已经谢顶. 同来的还有我们学校的同事俞书勤（访问学者）和周勇（研究生），我到他们家吃晚饭，他们两位为我找好了房子，地址是 First South, 1030 East, 离 Utah 大学数学系不远，但是有一个陡坡. 房东是一位美国老太太，叫做 Mrs. Morris, 九十多岁，她一个人住，十个房客都是中国学生或学者，相处非常友好. 我的房间在二层，宽敞明亮，房租每月 80 美元，没有涨过. 每月交给大使馆 20 美元，作为合作医疗之用.

　　Barnhill 的 CAGD 课程，每周三次，每次 50 分钟，我每次必到. G. Farin 从德国拿到了博士学位，也在数学系，什么职位我不知道. Barnhill 和 Farin 关系很好，是"铁哥们儿". Farin 也参加听课，还有 F. Little, 是 Barnhill 的助手，也很亲密. 很多时候，Barnhill 不讲，由 Farin 主讲，他在旁边听. 特别是 Farin 讲到 Bézier 三角曲面，我非常感兴趣，学到了很多的东西.

　　Little 喜欢讲话，特别是 Barnhill 讲课时，他喜欢插嘴. 有一次 Barnhill 上课，他不断地插话，搞得 Barnhill 很生气，拿着讲稿离开了教室. Little 有点不知如何是好，他说："我来讲吧！"我以为 Little 大祸临头，殊不知第三天两个人又有讲有笑.

　　因为我的来到，Barnhill 为我在他家举行 Party, 参加的有数学系的教授. 他的文章或校样也拿给我看，我提出的意见一般都得到采纳. 有一次他的校样要马上复印了，让我看一看. 我发现一个三角形三条边的垂直平分线，没有交于一点，我说，这个图错了，应该重画，他将信将疑，说："真的吗？"

　　Barnhill 把他主编的 1974 年 CAGD 论文集送给我，并签了他的名字. 有一些数学上的想法，我写成英文希望征求他的意见，能够跟他一起发表文章更好. 遗憾的是，他永远这么忙碌，我的草稿一直在他的书桌上. 他与 Farin 和 Little 无所不谈，而我由于语言障碍，不能畅所欲言. 我到 Utah 大学已经一年四个月了，一篇文章也没有发出去，我有些着急了.

　　正好 Brown 大学应用数学系 P. J. Davis 教授在 Utah 大学数学系访问，是 Barnhill 教授请过来的. 我到他的宾馆时，第一次见到他本人. 在中国科大图书馆，我看过他的著作《Interpolation and Approximation》. 我向 Davis 表示到 Brown 大学当访问学者的意向，他非常欢迎，但是回到 Utah 大学之后，还有一些手续要办理.

我是 1981 年底离开 Utah 大学的. 从盐湖城到波士顿的飞机票是 Barnhill 的基金资助的. 1982 年元旦，我应邀拜会了 Davis 教授. 应用数学系是古堡式的建筑，非常别致. Davis 的办公室是在三楼西头，面积很大，放两张大书桌，还空空荡荡，我们面对面地办公、写字，互不干扰. 他很早到办公室，而我十点左右才到，有时候我们一起共进午餐，他下午两点左右离开学校回家，先打电话给 Davis 夫人："我回家了."我通常是在办公室工作到深夜. 那时候，Davis 已经不做数学研究，专门写他的高级科普. 这七个月是我工作效率比较高的时期，我与 Davis 先后发表了三篇论文，其中有一篇为《The Convexity of Bernstein Polynomials over Triangles》，1984 年 1 月发表在美国《逼近论杂志》上，受到了多维逼近论和 CAGD 界的重视、关注、引用和应用. 我的三篇文章，经过他的细心修改，确实增色不少. 这是我第一次与大牌教授联名发表文章，也是一次新鲜经历.

这篇文章主要的结果有两条：第一个是证明了"三角域中，如果 Bernstein-Bézier 网是凸的，那么对应的 Bernstein-Bézier 曲面一定是凸的". 第二个是证明了三角域上 Bernstein-Bézier 曲面的分割定理. 这两个定理在理论上和实践上都非常重要. 图示如下：

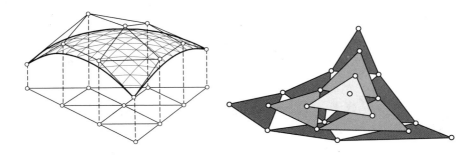

1982 年 7 月 25 日，在《人民日报》（海外版）中，刊登了我所写的短文，叫《看小松鼠有感》，描写 Davis 教授带我去乡村的经过，居然发表了. 我把我的文章翻译成英文，Davis 非常高兴. 8 月份我即将回到祖国，特此奉献给尊敬的 Davis 教授.

看小松鼠有感

我是在美国布朗大学工作的中国访问学者. 旅美期间, 有点感受很想和祖国读者谈谈.

布朗大学所在地普罗维登斯, 是美国东海岸罗得岛的首府. 这里经常可以看到这种情景: 有些小松鼠, 翘着银灰色的长尾巴, 一蹦一跳地横穿马路, 或者在树丛中爬来爬去. 它们同人们和谐共处, 相安无事. 在很长一段时间里, 我从未见过成年人或孩子去伤害这些野生小动物.

另一件事也给我留下深刻的印象. 那是一个久雨之后的晴天, 戴维斯教授开车陪我去康涅狄格州的一个乡村游览. 他今年五十九岁了, 是应用数学和计算数学方面的知名学者. 汽车在公路上疾驰, 两旁连绵不断的茂密树林, 随山起伏, 层层叠叠, 一片翠绿. 当汽车转入乡间小路后不久, 戴维斯教授突然紧急刹车, 我的视线随之集中到汽车前方. 原来, 小路正中站着一只小鸟, 也许是它因病残不及躲避吧. 老教授松了一口气说:"啊, 我差点儿杀死了它!" 接着, 他减慢速度, 小心翼翼地驾车让开小鸟, 保护了它. 老教授把车开过之后还不大放心, 要我回头看看小鸟是否还活在地上. 当他听说小鸟安然无恙时, 才微笑着踩大油门, 汽车又飞快地奔驰起来.

从我所亲身经历的这些小事, 使我看到爱护花草树木, 保护野生动物, 在这里已成为人们日常生活中的习惯. 正是这种良好的习惯, 在一定程度上保护了自然界的生态平衡, 有利于改善人们的生活环境. 我写这封短信, 目的是想让我们的一些同志从这些小事中得到一些启示.

下面的文章原载于:《Journal of Approximation》, 1984 年, 40 卷第 11~28 页.

The Convexity of Bernstein Polynomials over Triangles

Geng-zhe Chang* [1] and Philip J. Davis

Division of Applied Mathematics, Brown University, Providence, Rhode Island 02912

Communicated by Oved Shisha

Received July 14, 1982

Abstract: A necessary and sufficient condition for the convexity of the Bernstein polynomial over the triangle is presented. In particular, it follows that if the nth Bézier net of the function is convex over the triangle, so is the nth Bernstein polynomial.

1 Introduction

For a function $f(x)$ defined in $[0,1]$, the nth Bernstein polynomial of f is denoted by $B_n(f;x)$. It is well known that (see [1])

(1) if $f(x)$ is convex in $[0,1]$, so is $B_n(f;x)$;

(2) if $f(x)$ is convex in $[0,1]$, then

$$B_n(f;x) \geqslant B_{n+1}(f;x), \quad n=1,2,3,\cdots,$$

for $x \in [0,1]$.

We consider the possibility of extending these results to the Bernstein polynomials over triangles. Let us begin with some definitions and notation.

Let T_1, T_2, T_3 be three vertices of a triangle T which is called the base triangle. It is known that every point P of the plane in which the triangle lies

---

[1] On leave from the Department of Mathematics, University of Science and Technology of China, Hefei, P. R. of China.

can be expressed uniquely by $P = uT_1 + vT_2 + wT_3$ such that

$$u + v + w = 1. \tag{1}$$

(u, v, w) are called the barycentric coordinates of P with respect to the triangle T. We identify the point P with its barycentric coordinates and write $P = (u, v, w)$. It is clear that $T_1 = (1, 0, 0)$, $T_2 = (0, 1, 0)$, and $T_3 = (0, 0, 1)$.

Barycentric coordinates of points inside or on the boundary of T are characterized by (1) and

$$u \geqslant 0, \quad v \geqslant 0, \quad w \geqslant 0. \tag{2}$$

A function $f(P)$ defined on T can be expressed in terms of the barycentric coordinates of P, i.e., $f(P) = f(u, v, w)$. We compute $\dfrac{(n+1)(n+2)}{2}$ functional values of f:

$$f_{i,j,k} \equiv f\left(\frac{i}{n}, \frac{j}{n}, \frac{k}{n}\right), \quad i \geqslant 0, j \geqslant 0, k \geqslant 0, i+j+k = n.$$

The nth Bernstein polynomial of f over T is given by

$$B_n(f; P) = \sum_{i+j+k=n} f_{i,j,k} J^n_{i,j,k}(P), \tag{3}$$

where

$$J^n_{i,j,k}(P) = \frac{n!}{i!j!k!} u^i v^j w^k \tag{4}$$

are called the Bernstein basis polynomials.

Let Ω be a convex set in the plane. A continuous function $f(P)$ is said to be convex in Ω if

$$f\left(\frac{P+Q}{2}\right) \leqslant \frac{1}{2}[f(P) + f(Q)]$$

for all points P and Q in Ω.

As we tried to extend (1) and (2) to the Bernstein polynomials over triangles, we found that (2) can be extended while (1) cannot! For example, $f(P)$ is defined by the shaded triangles over T (see Fig 1), where $f(T_2) = 1$ and $f(T_1) = f(T_3) = f(M) = 0$ and where M is the midpoint of $\overline{T_2 T_3}$. It is clear that $f(P)$ is convex in T. Simple calculation shows that

$$B_2(f; P) = v(u + v),$$

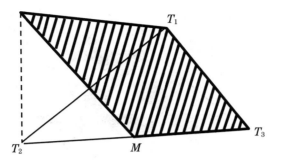

Fig 1

and that
$$B_2(f;T_1) = 0,$$
$$B_2(f;M) = B_2\left(f;0,\frac{1}{2},\frac{1}{2}\right) = \frac{1}{4}.$$

Since
$$B_2\left(f;\frac{1}{2}(T_1+M)\right) = B_2\left(f;\frac{1}{2},\frac{1}{4},\frac{1}{4}\right) = \frac{3}{16} > \frac{2}{16} = \frac{1}{2}\left(0+\frac{1}{4}\right)$$
$$= \frac{1}{2}[B_2(f;T_1) + B_2(f;M)],$$

it follows that $B_2(f;P)$ is not convex!

In this paper, a simple condition which ensures the convexity of $B_2(f;P)$ is given. To formulate our main results, some notation and terminology are needed.

Setting $F_{i,j,k} \equiv (\frac{i}{n}, \frac{j}{n}, \frac{k}{n}; f_{i,j,k})$, this is a point on the surface associated with the function $f(P)$. There are altogether $\frac{(n+1)(n+2)}{2}$ such points in the space. Drawing a triangle with three points
$$F_{i+1,j,k}, F_{i,j+1,k}, F_{i,j,k+1}$$
as its vertices, where $i+j+k = n-1$, a piecewise linear function on T is obtained and is denoted by $\hat{f}_n(P)$. $\hat{f}_n(P)$ is called the nth Bézier net of $f(P)$, in accordance with literature in Computer Aided Geometric Design[2].

The projection of $\hat{f}_n(P)$ onto triangle T produces a subdivision of T denoted by $S_n(T)$. $S_4(T)$ is illustrated in Fig 2.

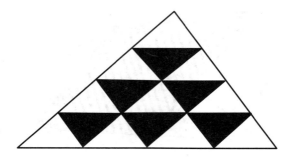

Fig 2

Our main results are the following:

(1) if the nth Bézier net $\hat{f}_n(P)$ is convex, so is the nth Bernstein polynomial $B_n(f;P)$;

(2) if $f(P)$ is convex in T, then we have

$$B_n(f;P) \geqslant B_{n+1}(f;P), \quad n=1,2,3,\cdots,$$

for $P \in T$.

2 Preliminaries

It is clear that

$$B_n(f;1,0,0) = f(1,0,0),$$
$$B_n(f;0,1,0) = f(0,1,0),$$
$$B_n(f;0,0,1) = f(0,0,1),$$

i.e., $B_n(f;P)$ interpolates to function f at the vertices of the base triangle T.

Since

$$J_{i,j,k}^n(P) \geqslant 0 \quad \text{for} \quad P \in T,$$

and

$$\sum_{i+j+k=n} J_{i,j,k}^n(P) = (u+v+w)^n = 1,$$

it follows that $B_n(f;P)$ is a convex linear combination of $\{f_{i,j,k}\}$. This means that the surface over T represented by $B_n(f;P)$ is contained in the convex hull of the set of points $\{F_{i,j,k}\}$.

There is a recursive algorithm for the evaluation of $B_n(f;P)$ (see [2]). Define

$$f^0_{i,j,k}(P) = f_{i,j,k}, \quad i+j+k = n, \tag{5}$$

and

$$f^l_{i,j,k}(P) = u f^{l-1}_{i+1,j,k}(P) + v f^{l-1}_{i,j+1,k}(P) + w f^{l-1}_{i,j,k+1}(P), \tag{6}$$

where $l = 1, 2, \cdots, n; i+j+k+l = n$. Introducing three formal "partial shift" operators E_1, E_2, E_3 by

$$E_1 f_{i,j,k} = f_{i+1,j,k},$$
$$E_2 f_{i,j,k} = f_{i,j+1,k},$$
$$E_3 f_{i,j,k} = f_{i,j,k+1},$$

then (6) can be rewritten as

$$f^l_{i,j,k}(P) = (uE_1 + vE_2 + wE_3) f^{l-1}_{i,j,k}(P). \tag{7}$$

Using (7) repeatedly, we have

$$f^l_{i,j,k}(P) = (uE_1 + vE_2 + wE_3)^l f_{i,j,k}. \tag{8}$$

Since E_1, E_2, E_3 commute, we can expand $(uE_1 + vE_2 + wE_3)^l$ in (8) by the trinomial formula and get

$$f^l_{i,j,k}(P) = \sum_{r+s+t=l} \frac{l!}{r!s!t!} u^r v^s w^t E_1^r E_2^s E_3^t f_{i,j,k},$$

i.e.

$$f^l_{i,j,k}(P) = \sum_{r+s+t=l} J^l_{r,s,t}(P) f_{i+r,j+s,k+t}, \tag{9}$$

where $i+j+k+l = n$. Putting $l = n$ in (9) we obtain

$$f^n_{0,0,0}(P) = B_n(f;P), \tag{10}$$

(10) implies that (6) together with (5) provides a stable recursive algorithm for evaluating the nth Bernstein polynomial over triangles.

Replacing $f_{i,j,k}$ by $F_{i,j,k}$ in both (5) and (6), we will have a recursive algorithm for determining the point on the Bernstein triangular surface, i.e.,

$$F_{0,0,0}^n(P) = [P; B_n(f;P)]. \tag{11}$$

We shall prove in the next section that the following three points

$$F_{1,0,0}^{n-1}(P) = \left[\frac{1+(n-1)u}{n}, \frac{(n-1)v}{n}, \frac{(n-1)w}{n}; f_{1,0,0}^{n-1}(P)\right],$$
$$F_{0,1,0}^{n-1}(P) = \left[\frac{(n-1)u}{n}, \frac{1+(n-1)v}{n}, \frac{(n-1)w}{n}; f_{0,1,0}^{n-1}(P)\right],$$
$$F_{0,0,1}^{n-1}(P) = \left[\frac{(n-1)u}{n}, \frac{(n-1)v}{n}, \frac{1+(n-1)w}{n}; f_{0,0,1}^{n-1}(P)\right],$$

determine a plane which is tangential to the Bernstein surface at the point $F_{0,0,0}^n(P)$.

Fig 3 shows the construction for $B_n\left(f; \frac{1}{2}, \frac{1}{4}, \frac{1}{4}\right)$.

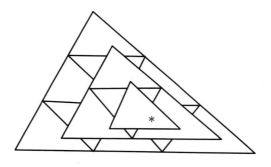

Fig 3

3 The Taylor expansion

We mention that not only $f_{0,0,0}^n(P)$, but also all the other $f_{i,j,k}^l(P)$, can be directly related to the Bernstein polynomial $B_n(f;P)$. We have the following

Lemma 1 *For $l = 0, 1, 2, \cdots, n$ and $i + j + k = l$ we have*

$$f_{i,j,k}^{n-l}(P) = \frac{(n-l)!}{n!} \frac{\partial^l}{\partial u^i \partial v^j \partial w^k} B_n(f;P), \tag{12}$$

where u, v, w are treated as independent variables.

Proof Since
$$\frac{\partial^l}{\partial u^i \partial v^j \partial w^k}(uE_1 + vE_2 + wE_3)^n$$
$$= \frac{n!}{(n-l)!}(uE_1 + vE_2 + wE_3)^{n-l} E_1^i E_2^j E_3^k,$$

we have by (8),
$$\frac{\partial^l}{\partial u^i \partial v^j \partial w^k} B_n(f;P) = \frac{\partial^l}{\partial u^i \partial v^j \partial w^k}(uE_1 + vE_2 + wE_3)^n f_{0,0,0}$$
$$= \frac{n!}{(n-l)!}(uE_1 + vE_2 + wE_3)^{n-l} f_{i,j,k}$$
$$= \frac{n!}{(n-l)!} f_{i,j,k}^{n-l}(P).$$

Equation (10) is a special case of (12) in which $l = 0$. □

We are now in a position to present the Taylor expansion for the Bernstein polynomial $B_n(f;P)$. Put $P = (u,v,w)$, $P' = (u',u',w')$ and $P' - P = (u' - u, v' - v, w' - w)$, then we have

Theorem 2 *For any P and P', we have the identity*
$$B_n(f;P') = \sum_{l=0}^n \binom{n}{l} \sum_{i+j+k=l} f_{i,j,k}^{n-l}(P) J_{i,j,k}^l(P' - P). \tag{13}$$

Proof *Using the Taylor expansion for polynomials of degree n in three variables, we get*
$$B_n(f;P') = \sum_{l=0}^n \frac{1}{l!} \left[(u-u')\frac{\partial}{\partial u} + (v'-v)\frac{\partial}{\partial v} + (w'-w)\frac{\partial}{\partial w}\right]^l B_n(f;P)$$
$$= \sum_{l=0}^n \frac{1}{l!} \sum_{i+j+k=l} \frac{l!}{i!j!k!}(u'-u)^i(v'-v)^j(w'-w)^k \frac{\partial^l B_n(f;P)}{\partial u^i \partial v^j \partial w^k}$$
$$= \sum_{l=0}^n \frac{1}{l!} \sum_{i+j+k=l} J_{i,j,k}^l(P'-P) \frac{\partial^l B_n(f;P)}{\partial u^i \partial v^j \partial w^k},$$

and by (12) we have
$$B_n(f;P') = \sum_{l=0}^n \frac{n!}{l!(n-l)!} \sum_{i+j+k=l} f_{i,j,k}^{n-l}(P) J_{i,j,k}^l(P'-P).$$

This completes the proof of Theorem 2. □

The Taylor expansion of functions (13) provides a powerful tool for the investigation of the local analytical behavior of the Bernstein polynomial in the neighborhood of P. Let us write the first three terms of the right-hand side of (13) in more detail:

$$\begin{aligned}B_n(f;P') =& f^n_{0,0,0}(P)+n[f^{n-1}_{1,0,0}(P)(u'-u)\\&+f^{n-1}_{0,1,0}(P)(v'-v)+f^{n-1}_{0,0,1}(P)(w'-w)]\\&+\frac{n(n-1)}{2}[u'-u,v'-v,w'-w]\\&\times\begin{bmatrix}f^{n-2}_{2,0,0} & f^{n-2}_{1,1,0} & f^{n-2}_{1,0,1}\\ f^{n-2}_{1,1,0} & f^{n-2}_{0,2,0} & f^{n-2}_{0,1,1}\\ f^{n-2}_{1,0,1} & f^{n-2}_{0,1,1} & f^{n-2}_{0,0,2}\end{bmatrix}\begin{bmatrix}u'-u\\v'-v\\w'-w\end{bmatrix}\\&+\cdots.\end{aligned} \quad (14)$$

Note that the elements in the 3×3 matrix should be evaluated at the point P.

The first four terms in the right-hand side of (14) form a linear function in u', v', w', which has the contact of at least second degree with the surface at the point $F^n_{0,0,0}(P)$. Hence

$$\begin{aligned}z =& f^n_{0,0,0}(P)+n\Big[f^{n-1}_{1,0,0}(P)(u'-u)\\&+f^{n-1}_{0,1,0}(P)(v'-v)+f^{n-1}_{0,0,1}(P)(w'-w)\Big]\end{aligned} \quad (15)$$

is the tangent plane of the Bernstein surface at the point $F^n_{0,0,0}(P)$. It is easy to show that the plane determined by three points $F^{n-1}_{1,0,0}(P)$, $F^{n-1}_{0,1,0}(P)$, $F^{n-1}_{0,0,1}(P)$, also has Eq(15). Thus the conclusion in the end of the previous section is justified.

If $f^{n-1}_{1,0,0}(P)$, $f^{n-1}_{0,1,0}(P)$, $f^{n-1}_{0,0,1}(P)$ are not all equal, then without loss of generality we can assume that $f^{n-1}_{1,0,0}(P)\neq f^{n-1}_{0,1,0}(P)$. In this case we put $w'=w$ and since $u'-u=-(v'-v)$,

$$\begin{aligned}&f^{n-1}_{1,0,0}(P)(u'-u)+f^{n-1}_{0,1,0}(P)(v'-v)+f^{n-1}_{0,0,1}(P)(w'-w)\\&=[f^{n-1}_{1,0,0}(P)-f^{n-1}_{0,1,0}(P)](u'-u),\end{aligned}$$

and this will assume both positive and negative values no matter how small $|u'-u|$ is. Thus we have

Theorem 3 For $f_{0,0,0}^n(P)$ to be a local extreme value it is necessary that

$$f_{1,0,0}^{n-1}(P) = f_{0,1,0}^{n-1}(P) = f_{0,0,1}^{n-1}(P).$$

This means geometrically that the tangent plane of the Bernstein surface at the point $F_{0,0,0}^n(P)$ must be parallel to the plane determined by the base triangle. To determine whether $f_{0,0,0}^n$ is a local extreme value or not, we need further information coming from the third term in the right-hand side of (14), i.e., from the following quadratic form

$$\begin{bmatrix} \xi & \eta & \zeta \end{bmatrix} \begin{bmatrix} f_{2,0,0}^{n-2} & f_{1,1,0}^{n-2} & f_{1,0,1}^{n-2} \\ f_{1,1,0}^{n-2} & f_{0,2,0}^{n-2} & f_{0,1,1}^{n-2} \\ f_{1,0,1}^{n-2} & f_{0,1,1}^{n-2} & f_{0,0,2}^{n-2} \end{bmatrix} \begin{bmatrix} \xi \\ \eta \\ \zeta \end{bmatrix},$$

where $\xi + \eta + \zeta = (u'+v'+w') - (u+v+w) = 1-1 = 0$. This quadratic form will be studied carefully in the next section.

4 Convexity

On p.80, Sect.99 of the book[3], the investigation of convexity of a function Φ, with the rectangular Cartesian coordinates as its variables, is shifted to that of nonnegativity of the quadratic form with the second order partial derivatives of Φ as its coefficients. A necessary and sufficient condition for the convexity of Φ is presented there. With obvious modifications we can state

Theorem 4 Ω *is a convex set in the plane. A necessary and sufficient condition that $B_n(f;P)$ should be convex in Ω is that the quadratic form* (15) *should be nonnegative for all P in Ω and all (ξ, η, ζ) such that $\xi + \eta + \zeta = 0$.*

Setting for simplicity

$$A = f_{2,0,0}^{n-2}(P), \quad B = f_{0,2,0}^{n-2}(P), \quad C = f_{0,0,2}^{n-2}(P),$$
$$a = f_{0,1,1}^{n-2}(P), \quad b = f_{1,0,1}^{n-2}(P), \quad c = f_{1,1,0}^{n-2}(P),$$

Eq(15) becomes

$$\begin{bmatrix} \xi & \eta & \zeta \end{bmatrix} \begin{bmatrix} A & c & b \\ c & B & a \\ b & a & C \end{bmatrix} \begin{bmatrix} \xi \\ \eta \\ \zeta \end{bmatrix}, \qquad (16)$$

where $\xi + \eta + \zeta = 0$. Insertion of $\zeta = -\xi - \eta$ in (16) gives

$$\begin{bmatrix} \xi & \eta \end{bmatrix} \begin{bmatrix} A+C-2b & C+c-a-b \\ C+c-a-b & B+C-2a \end{bmatrix} \begin{bmatrix} \xi \\ \eta \end{bmatrix}, \qquad (17)$$

where now there are no longer any restrictions on ξ and η. Note that the quadratic form (17) is nonnegative if and only if

$$\begin{aligned} & A+C \geqslant 2b, \qquad B+C \geqslant 2a, \\ & (A+C-2b)(B+C-2a) \geqslant (C+c-a-b)^2. \end{aligned} \qquad (18)$$

The second inequality is equivalent to

$$\begin{aligned} BC + CA + AB &+ 2(bc+ca+ab) \\ &\geqslant a^2 + b^2 + c^2 + 2(Aa+Bb+Cc). \end{aligned} \qquad (19)$$

Hence Theorem 4 can be reformulated by

Theorem 4′ *A necessary and sufficient condition that $B_n(f;P)$ should be convex in Ω is that* (18) *and* (19) *hold for all P in Ω.*

The following theorem provides a sufficient condition for the convexity of $B_n(f;P)$ in Ω. This condition is easier to check.

Theorem 5 *If for all P in Ω we have that*

$$A + a \geqslant b + c, \qquad (20)$$

$$B + b \geqslant c + a, \qquad (21)$$

$$C + c \geqslant a + b, \qquad (22)$$

then $B_n(f;P)$ is convex in Ω.

Proof It is clear that (20), (21), (22) imply

$$A \geqslant b + c - a, \qquad (23)$$

$$B \geqslant c+a-b, \tag{24}$$

$$C \geqslant a+b-c, \tag{25}$$

and that

$$\frac{1}{2}(B+C) - a, \tag{26}$$

$$\frac{1}{2}(C+A) - b, \tag{27}$$

$$\frac{1}{2}(A+B) - c, \tag{28}$$

are nonnegative numbers. Multiplying both sides of (23), (24), (25) by the numbers in (26), (27), (28), respectively, adding, and simplifying, we get (19). The nonnegativity of numbers in (26) and (27) implies (18). Hence Theorem 5 comes from Theorem 4'. □

It will be desirable if we can find some conditions for convexity of $B_n(f;P)$ in terms of $f_{i,j,k}$ the values of the primitive function.

The set $\{f_{i,j,k}; i+j+k = n\}$ is said to be convex in the u-direction if inequalities

$$f_{i+1,j,k} + f_{i-1,j+1,k+1} \geqslant f_{i,j+1,k} + f_{i,j,k+1} \tag{29}$$

hold for all i, j, k such that $i > 0$ and $i+j+k = n-1$. Let us say a few words about inequality (29). In the subdivision $S_n(T)$ there are altogether $\frac{n(n-1)}{2}$ parallelograms each of which has the diagonal parallel to the side $u = 0$ of the base triangle. A typical parallelogram with its vertices $\left(\frac{(i+1)}{n}, \frac{j}{n}, \frac{k}{n}\right)$, $\left(\frac{i}{n}, \frac{(j+1)}{n}, \frac{k}{n}\right)$, $\left(\frac{(i-1)}{n}, \frac{(j+1)}{n}, \frac{(k+1)}{n}\right)$, $\left(\frac{i}{n}, \frac{j}{n}, \frac{(k+1)}{n}\right)$ and the valuations of the function f at these vertices are shown in Fig 4. Inequality (29) has the following interpretations: in each of these parallelograms the sum of values of f at two vertices connected by the explicit diagonal is less than or equal to that of values of f at other two vertices. Similar definitions may be applied to the v-direction and the w-direction by

$$f_{i,j+1,k} + f_{i+1,j-1,k+1} \geqslant f_{i+1,j,k} + f_{i,j,k+1}, \quad j > 0 \text{ and } i+j+k = n-1, \tag{30}$$

$$f_{i,j,k+1} + f_{i+1,j+1,k-1} \geqslant f_{i+1,j,k} + f_{i,j+1,k}, \quad k > 0 \text{ and } i+j+k = n-1, \tag{31}$$

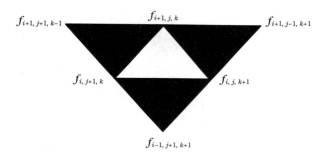

Fig 4

respectively (see Fig 4). By the recursive algorithm (6) we have

$$f^1_{i+2,j-1,k-1}(P) = uf_{i+3,j-1,k-1} + vf_{i+2,j,k-1} + wf_{i+2,j-1,k},$$
$$f^1_{i,j,k}(P) = uf_{i+1,j,k} + vf_{i,j+1,k} + wf_{i,j,k+1},$$

and

$$f^1_{i+1,j,k-1}(P) = uf_{i+2,j,k-1} + vf_{i+1,j+1,k-1} + wf_{i+1,j,k},$$

$$f^1_{i+1,j-1,k}(P) = uf_{i+2,j-1,k} + vf_{i+1,j,k} + wf_{i+1,j-1,k+1}.$$

Inequalities (29) imply that for $u \geqslant 0$, $v \geqslant 0$, $w \geqslant 0$,

$$f^1_{i+2,j-1,k-1}(P) + f^1_{i,j,k}(P) \geqslant f^1_{i+1,j,k-1}(P) + f^1_{i+1,j-1,k}(P).$$

In other words, the convexity of the set $\{f^1_{i,j,k}\}$ in the u-direction implies the convexity of the set $\{f^1_{i,j,k}(P)\}$ in the u-direction for P inside the base triangle T. Repeating this argument we conclude that the convexity of the set $\{f_{i,j,k}(P)\}$ in the u-direction implies the convexity of $\{f^{n-2}_{i,j,k}(P)\}$ in the u-direction for $P \in T$, or equivalently implies the inequality (20). Similar reasoning can be applied for the convexity in the v- direction and in the w-direction. Hence inequalities in (29), (30), (31) imply inequalities (20), (21), (22) for $P \in T$. Thus we have

Theorem 6 *If $\{f_{i,j,k}\}$ satisfy inequalities* (29), (30), (31), *then $B_n(f;P)$ is*

convex over the base triangle T.

Corollary 7 *If the nth Bézier net $\hat{f}_n(P)$ is convex in T, so is $B_n(f;P)$.*

Proof In this case we have

$$f_{i,j+1,k} + f_{i,j,k+1} = f\left(\frac{i}{n}, \frac{j+1}{n}, \frac{k}{n}\right) + f\left(\frac{i}{n}, \frac{j}{n}, \frac{k+1}{n}\right)$$
$$= \hat{f}_n\left(\frac{i}{n}, \frac{j+1}{n}, \frac{k}{n}\right) + \hat{f}_n\left(\frac{i}{n}, \frac{j}{n}, \frac{k+1}{n}\right).$$

By the definition of $\hat{f}_n$, $\hat{f}_n$ is linear along the line segment between points $\left(\frac{i}{n}, \frac{j+1}{n}, \frac{k}{n}\right)$ and $\left(\frac{i}{n}, \frac{j}{n}, \frac{k+1}{n}\right)$, hence the value of $\hat{f}_n$ at the midpoint of the segment is half of the sum of the values of $\hat{f}_n$ at two endpoints. Thus we have

$$f_{i,j+1,k} + f_{i,j,k+1} = 2\hat{f}_n\left(\frac{i}{n}, \frac{j+\frac{1}{2}}{n}, \frac{k+\frac{1}{2}}{n}\right),$$

and by the convexity of $\hat{f}_n$,

$$2\hat{f}_n\left(\frac{i}{n}, \frac{j+\frac{1}{2}}{n}, \frac{k+\frac{1}{2}}{n}\right)$$
$$\leqslant \hat{f}_n\left(\frac{i+1}{n}, \frac{j}{n}, \frac{k}{n}\right) + \hat{f}_n\left(\frac{i-1}{n}, \frac{j+1}{n}, \frac{k+1}{n}\right)$$
$$= f\left(\frac{i+1}{n}, \frac{j}{n}, \frac{k}{n}\right) + f\left(\frac{i-1}{n}, \frac{j+1}{n}, \frac{k+1}{n}\right)$$
$$= f_{i+1,j,k} + f_{i-1,j+1,k+1},$$

as the point $\left(\frac{i}{n}, \frac{j+\frac{1}{2}}{n}, \frac{k+\frac{1}{2}}{n}\right)$ is the midpoint of the line segment between points $\left(\frac{i+1}{n}, \frac{j}{n}, \frac{k}{n}\right)$ and $\left(\frac{i-1}{n}, \frac{j+1}{n}, \frac{k+1}{n}\right)$ too. Hence we get

$$f_{i,j+1,k} + f_{i,j,k+1} \leqslant f_{i+1,j,k} + f_{i-1,j+1,k+1},$$

for all i, j, k such that $i > 0$ and $i+j+k = n-1$. This inequality is just (29). Hence the convexity of $\hat{f}_n(P)$ in T implies (29), (30) and (31). By Theorem 5 we conclude that $B_n(f;P)$ is convex in the base triangle T. □

5 Condition for $B_n(f;P) = B_{n+1}(f;P)$

If $f(P)$ is continuous in Ω, then the convexity of f in Ω can be defined equivalently by (see [3])

$$f\left(\sum_{k=1}^{m} \lambda_k P_k\right) \leqslant \sum_{k=1}^{m} \lambda_k f(P_k) \tag{32}$$

for any $P_1, P_2, \cdots, P_m$ in Ω and for any nonnegative numbers $\lambda_1, \lambda_2, \cdots, \lambda_m$ such that

$$\lambda_1 + \lambda_2 + \cdots + \lambda_m = 1. \tag{33}$$

Lemma 8 *We have the identity*

$$J_{i,j,k}^n(P) = \frac{1}{n+1}[(i+1)J_{i+1,j,k}^{n+1}(P) + (j+1)J_{i,j+1,k}^{n+1}(P) \\ + (k+1)J_{i,j,k+1}^{n+1}(P)], \tag{34}$$

where $i+j+k=n$.

This lemma can be verified by simple calculations. Eq (34) enables us to write the nth Bernstein polynomial $B_n(f;P)$ in terms of the $(n+1)$th Bernstein basis polynomials:

$$\begin{aligned}B_n(f;P) = &\sum_{i+j+k=n} \frac{i+1}{n+1} f\left(\frac{i}{n},\frac{j}{n},\frac{k}{n}\right) J_{i+1,j,k}^{n+1}(P) \\ &+ \sum_{i+j+k=n} \frac{j+1}{n+1} f\left(\frac{i}{n},\frac{j}{n},\frac{k}{n}\right) J_{i,j+1,k}^{n+1}(P) \\ &+ \sum_{i+j+k=n} \frac{k+1}{n+1} f\left(\frac{i}{n},\frac{j}{n},\frac{k}{n}\right) J_{i,j,k+1}^{n+1}(P)\end{aligned} \tag{35}$$

Replacing $(i+1)$ by i, the first term of the right-hand side of (35) becomes

$$\sum_{i+j+k=n+1} \frac{i}{n+1} f\left(\frac{i-1}{n},\frac{j}{n},\frac{k}{n}\right) J_{i,j,k}^{n+1}(P).$$

Even though $f(\frac{i-1}{n},\frac{j}{n},\frac{k}{n})$ makes no sense for $i=0$, the coefficient $\frac{i}{n+1}$ standing before f will annihilate the corresponding term. Applying similar manipulations to the second and the third term in the right-hand side of (35), we obtain

$$B_n(f;P) = \sum_{i+j+k=n+1} \frac{1}{n+1} \left[if\left(\frac{i-1}{n}, \frac{j}{n}, \frac{k}{n}\right) \right. \\ \left. + jf\left(\frac{i}{n}, \frac{j-1}{n}, \frac{k}{n}\right) + kf\left(\frac{i}{n}, \frac{j}{n}, \frac{k-1}{n}\right) \right] J_{i,j,k}^{n+1}(P). \tag{36}$$

If $f(P)$ is convex and continuous in T, since $i+j+k=n+1$ and

$$\frac{i}{n+1}\left(\frac{i-1}{n}, \frac{j}{n}, \frac{k}{n}\right) + \frac{j}{n+1}\left(\frac{i}{n}, \frac{j-1}{n}, \frac{k}{n}\right) + \frac{k}{n+1}\left(\frac{i}{n}, \frac{j}{n}, \frac{k-1}{n}\right) \\ = \left(\frac{i}{n+1}, \frac{j}{n+1}, \frac{k}{n+1}\right), \tag{37}$$

then by (32) we get

$$\frac{1}{n+1}\left[if\left(\frac{i-1}{n}, \frac{j}{n}, \frac{k}{n}\right) + jf\left(\frac{i}{n}, \frac{j-1}{n}, \frac{k}{n}\right) + kf\left(\frac{i}{n}, \frac{j}{n}, \frac{k-1}{n}\right) \right] \\ \geqslant f\left(\frac{i}{n+1}, \frac{j}{n+1}, \frac{k}{n+1}\right). \tag{38}$$

By (36) and (38) we see that if the continuous function $f(P)$ is convex in T, then

$$B_n(f;P) \geqslant B_{n+1}(f;P) \tag{39}$$

for all $P \in T$ and $n = 1, 2, 3, \cdots$. We propose the following problem: Under what conditions does the equality in (39) hold? From (36) and the linear independence of $J_{i,j,k}^{n+1}(P)$ we see that for any function $f(P)$ (not necessarily convex) $B_n(f;P) \equiv B_{n+1}(f;P)$ if and only if

$$\frac{1}{n+1}\left[if\left(\frac{i-1}{n}, \frac{j}{n}, \frac{k}{n}\right) + jf\left(\frac{i}{n}, \frac{j-1}{n}, \frac{k}{n}\right) + kf\left(\frac{i}{n}, \frac{j}{n}, \frac{k-1}{n}\right) \right] \\ = f\left(\frac{i}{n+1}, \frac{j}{n+1}, \frac{k}{n+1}\right), \tag{40}$$

where $i+j+k = n+1$. If we call each point $F_{i,j,k}$ the vertex of the nth Bézier net $\hat{f}_n(P)$, we can state (40) geometrically as the following

Theorem 9 *Let f be any function defined in T. Then $B_n(f;P) \equiv B_{n+1}(f;P)$ if and only if all vertices of $\hat{f}_{n+1}(P)$ lie on $\hat{f}_n(P)$.*

For a convex function f, we can reformulate Theorem 9 in a different way which has a stronger geometric implication. We have mentioned in Section 1 that the projection of $f_n(P)$ onto the triangle T produces the subdivision of T denoted by $S_n(T)$. Each point $(\frac{i}{n}, \frac{j}{n}, \frac{k}{n})$ with $i+j+k=n$ is called a node of $S_n(T)$. $S_n(T)$ has $\frac{(n+1)(n+2)}{2}$ nodes altogether. Denote the boundary of T by ∂T and $T^\circ \equiv \frac{T}{\partial T}$. The nodes in T° are called interior nodes while the others are called boundary nodes. Clearly $S_n(T)$ has $3n$ boundary nodes and $\frac{(n-2)(n-1)}{2}$ interior nodes. There are n^2 subtriangles in $S_n(T)$. Each subtriangle with vertices

$$\left(\frac{i-1}{n}, \frac{j}{n}, \frac{k}{n}\right), \left(\frac{i}{n}, \frac{j-1}{n}, \frac{k}{n}\right), \left(\frac{i}{n}, \frac{j}{n}, \frac{k-1}{n}\right), \tag{41}$$

where $i,j,k \geqslant 1$ and $i+j+k=n+1$, is called a downward subtriangle. $S_n(T)$ has $\frac{(n-1)n}{2}$ downward subtriangles. All downward subtriangles of $S_4(T)$ are colored by black in Fig 2.

Let us observe the relationship between $S_n(T)$ and $S_{n+1}(T)$. All nodes of $S_{n+1}(T)$ can be put into three categories:

(1) Interior nodes are characterized by $(\frac{i}{n+1}, \frac{j}{n+1}, \frac{k}{n+1})$ with i,j, $k \geqslant 1$ and $i+j+k=n+1$. From (37) and (41) we see that each interior node of $S_{n+1}(T)$ lies inside one and only one downward triangle of $S_n(T)$. See Fig 5.

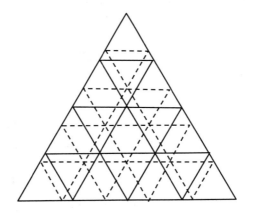

Fig 5

(2) Nodes on just one side of T are characterized by $(\frac{i}{n+1}, \frac{j}{n+1}, \frac{k}{n+1})$ with only one of i, j, k equal to zero. From (37) we can say that each of these nodes of $S_{n+1}(T)$ lies inside one and only one boundary segment of $S_n(T)$.

(3) Three vertices of T.

If $f(P)$ is convex in T, naturally $f(P)$ is convex in each subtriangle with vertices shown in (41). Hence (40) will imply $f(P)$ is linear over this subtriangle. Thus we have

Theorem 10 *Let function $f(P)$ be convex and continuous in T. Let D_n be the union of all downward subtriangles of $S_n(T)$. Then $B_{n+1}(f;P) = B_n(f;P)$ if and only if*

$$f(P) = \hat{f}_n(P) \quad for \quad P \in D_n \cup \partial T, \tag{42}$$

otherwise we have $B_n(f;P) > B_{n+1}(f;P)$ for $P \in T^\circ$.

6 Convexity over a changed triangle

Even if $f(P)$ is defined on the base triangle T only, the Bernstein polynomial (3) is well defined in the whole plane. In some practical applications, we relax the restrictions (2) for more flexibility. Let $T^* = \Delta T_1^* T_2^* T_3^*$ be any triangle in the same plane of the triangle T. We are interested in the convexity of the Bernstein polynomial $B_n(f;P)$ restricted to T^*. Assume T_i^* has barycentric coordinates (u_i, v_i, w_i) with respect to T, $i = 1, 2, 3$. We define the following $\frac{(n+1)(n+2)}{2}$ numbers

$$\begin{aligned}
f^*_{i,j,k} = \sum_{r+s+t=i} \sum_{\alpha+\beta+\gamma=j} \sum_{\lambda+\mu+\nu=k} & J^i_{r,s,t}(u_1,v_1,w_1) \\
\times\, & J^j_{\alpha,\beta,\gamma}(u_2,v_2,w_2) J^k_{\lambda,\mu,\nu}(u_3,v_3,w_3) \\
\times\, & f_{r+\alpha+\lambda,\, s+\beta+\mu,\, t+\gamma+\nu},
\end{aligned} \tag{43}$$

or briefly,

$$\begin{aligned}f^*_{i,j,k} =& (u_1E_1 + v_1E_2 + w_1E_3)^i(u_2E_1 + v_2E_2 + w_2E_3)^j \\ & \times (u_3E_1 + v_3E_2 + w_3E_3)^k f_{0,0,0},\end{aligned}$$

where $i + j + k = n$. We have

Theorem 11 *Let (u, v, w) be the barycentric coordinates of P with respect to the triangle T^*, then the expression*

$$\sum_{i+j+k=n} f^*_{i,j,k} J^n_{i,j,k}(u,v,w) \tag{44}$$

represents the Bernstein polynomial $B_n(f; P)$ restricted to T^*.

Proof Setting $l = r + \alpha + \lambda$, $m = s + \beta + \mu$, $p = t + \gamma + \nu$, we have $l + m + p = i + j + k = n$ and

$$\begin{aligned} & J^n_{i,j,k}(u,v,w) J^i_{r,s,t}(u_1,v_1,w_1) J^j_{\alpha,\beta,\gamma}(u_2,v_2,w_2) J^k_{\lambda,\mu,\nu}(u_3,v_3,w_3) \\ &= \frac{n!}{i!j!k!} \cdot \frac{i!}{r!s!t!} \cdot \frac{j!}{\alpha!\beta!\gamma!} \cdot \frac{k!}{\lambda!\mu!\nu!} u^i v^j w^k u_1^r v_1^s w_1^t u_2^\alpha v_2^\beta w_2^\gamma u_3^\lambda v_3^\mu w_3^\nu \\ &= \frac{n!}{l!m!p!} \cdot \frac{l!}{r!\alpha!\lambda!} \cdot \frac{m!}{s!\beta!\mu!} \cdot \frac{p!}{t!\gamma!\nu!} \\ & \quad \times (uu_1)^r (vu_2)^\alpha (wu_3)^\lambda \cdot (uv_1)^s (vv_2)^\beta (wv_3)^\mu (uw_1)^t (vw_2)^\gamma (ww_3)^\nu \\ &= \frac{n!}{l!m!p!} J^l_{r,\alpha,\lambda}(uu_1, vu_2, wu_3) \\ & \quad \times J^m_{s,\beta,\mu}(uv_1, vv_2, wv_3) \times J^p_{t,\gamma,\nu}(uw_1, vw_2, ww_3). \end{aligned}$$

Insertion of (43) into (44) gives

$$\begin{aligned} & \sum_{i+j+k=n} f^*_{i,j,k}(u,v,w) \\ &= \sum_{l+m+p=n} \frac{n!}{l!m!p!} \left[\sum_{r+\alpha+\lambda=l} J^l_{r,\alpha,\lambda}(uu_1, vu_2, wu_3) \right] \\ & \quad \cdot \left[\sum_{s+\beta+\mu=m} J^m_{s,\beta,\mu}(uv_1, vv_2, wv_3) \right] \\ & \quad \cdot \left[\sum_{t+\gamma+\nu=p} J^p_{t,\gamma,\nu}(uw_1, vw_2, ww_3) \right] f_{l,m,p} \end{aligned}$$

$$= \sum_{l+m+p=n} \frac{n!}{l!m!p!} (uu_1 + vu_2 + wu_3)^l.$$
$$(uv_1 + vv_2 + wv_3)^m (uw_1 + vw_2 + ww_3)^p f_{l,m,p}$$
$$= B_n(f; uu_1 + vu_2 + wu_3, uv_1 + vv_2 + wv_3, uw_1 + vw_2 + ww_3).$$

Note that
$$\begin{bmatrix} u & v & w \end{bmatrix} \begin{bmatrix} u_1 & v_1 & w_1 \\ u_2 & v_2 & w_2 \\ u_3 & v_3 & w_3 \end{bmatrix}$$

are the barycentric coordinates of a point inside T^* with respect to the triangle T. □

Now we can use the methods presented in Section 4 on $f_{i,j,k}^*$, to check the convexity of $B_n(f; P)$ restricted to the triangle T^*.

References

[1] Davis P J. Interpolation and Approximation[M]. New York: Dover, 1975.

[2] Barnhill R E, Farin G. C^1 Quintic Interpolation over Triangles: Two Explicit Representations[J]. Internat. J. Numer. Methods Engrg., 1981, 17:1763-1778.

[3] Hardy G H, Littlewood J E, Pólya G. Inequalities[M]. London: Cambridge Univ. Press, 1934.

[4] Farin G. Subsplines Ueber Dreicken [R]. West Germany: Bratmschweig, 1979.

[5] Farin G. Bézier Polynomials over Triangles and the Construction of Piecewise C^r Polynomials[D]. TR/91, Dept of Mathematics, Brunel University, Middlesex, 1980.

8

保凸性定理

    1981 年，冯玉瑜同志在美国做了两年访问学者后，学成回国. 他的主人是威斯康星大学数学研究所的 Carl de Boor 教授，著名的应用数学家、逼近论和样条函数专家. 1982 年，我也回到了中国科大. 在西方做访问学者的这两年中，收获成果显著，不仅学习了若干新课程，懂得了做科研的方法，还结识了西方的科学家，为以后国际交往铺平了道路.

    以我自己为例，1980 年我第一次出国，那时我已满 44 周岁，没有任何科研工作. 两年的时间，我有了显著进步. 感谢祖国和人民，用劳动人民的血汗钱，送我出国深造.

    冯玉瑜和我在数学系招收硕士研究生，用在美国学来的专长，为研究生讲课，我们也互相听课，共同主讲讨论班，毫无保留，亲密无间.

    我与 Davis 提出的凸性定理，虽然正确无误，但是其证明非常冗长，我记得其中还要构造一个不等式. 我与冯玉瑜在我们的论文《改进凸性定理》中用到了"移位算子"，使得证明简洁和流畅. 我们进一步证明，如果二次和三次 Berstein-Bézier 三角域曲面是凸的，那么二次和三次 Bézier 网也是凸的.

    对于一阶 Bézier 网，问题不证自明. 这时 Bézier 网就是一个三角形，曲面和网是重合的.

    考虑二次 Berstein-Bézier 三角曲面，应用重心坐标和矩阵，曲面可以

8 保凸性定理

写成

$$K = (u,v,w)\begin{pmatrix} A & c & b \\ c & B & a \\ b & a & C \end{pmatrix}\begin{pmatrix} u \\ v \\ w \end{pmatrix},$$

其中, 3×3 的方阵中的 6 个参数都是实数, $u \geqslant 0, v \geqslant 0, w \geqslant 0$ 是重心坐标并且 $u+v+w=1$. 上述 3 阶方阵可以分拆成

$$\begin{pmatrix} A & c & b \\ c & B & a \\ b & a & C \end{pmatrix} = \begin{pmatrix} A+a-b-c & 0 & 0 \\ 0 & B+b-c-a & 0 \\ 0 & 0 & C+c-a-b \end{pmatrix}$$
$$+ \begin{pmatrix} b+c-a & c & b \\ c & c+a-b & a \\ b & a & a+b-c \end{pmatrix},$$

最后的矩阵可以写成

$$(u,v,w)\begin{pmatrix} b+c-a & c & b \\ c & c+a-b & a \\ b & a & a+b-c \end{pmatrix}\begin{pmatrix} u \\ v \\ w \end{pmatrix}$$
$$= (b+c-a)u + (c+a-b)v + (a+b-c)w,$$

这里 u,v,w 都是 1 次的, 欲知详情可以参看 "国际数学奥林匹克竞赛" 一节.

二次 Berstein-Bézier 三角曲面是

$$K = (u,v,w)\begin{pmatrix} A & c & b \\ c & B & a \\ b & a & c \end{pmatrix}\begin{pmatrix} u \\ v \\ w \end{pmatrix}$$
$$= (A+a-b-c)u^2 + (B+b-c-a)v^2 + (C+c-a-b)w^2 + 线性项.$$

由于线性项对于 "凸" 和 "不凸" 没有影响, 只需要考虑 $(A+a-b-c)u^2 + (B+b-c-a)v^2 + (C+c-a-b)w^2$. 因为 u^2, v^2, w^2 都是凸函数, 二阶三角曲面是凸的当且仅当不等式

$$A+a-b-c \geqslant 0,$$
$$B+b-c-a \geqslant 0,$$

$$C+c-a-b\geqslant 0$$

均成立. 这三个不等式正好是 2 次 Bézier 网为凸的必要充分条件. 这就说明，2 次 Berstein-Bézier 曲面是凸曲面当且仅当相应的 Bézier 网是凸曲面.

对于 3 次 Berstein-Bézier 凸曲面，也有相应的充分必要条件.

1985 年，在《数学年刊》B 辑，第 6 卷 171~176 页，我与冯玉瑜合作，又一次推导了 Berstein-Bézier 三角域曲面的凸性. 由升阶算法，不断地迭代，最后生成凸曲面.

我与冯玉瑜和学生们，辛勤耕耘，日积月累. 我们的 CAGD 小组，不仅得到国内同行的公认，国外同行对于我们的工作也表示赞赏.

1992 年，中国科学院授予我们小组自然科学二等奖.

下面的文章原载于:《Computer Aided Geometric Design》, 1984 年, 1 卷第 279~283 页.

An Improved Condition for the Convexity of Bernstein-Bézier Surfaces over Triangles

Geng-zhe Chang and Yu-yu Feng

Department of Mathematics, University of Science and Technology of China,
Hefei, Anhui, The People's Republic of China

Abstract: A sufficient condition which is superior to that of Chang and Davis for the convexity of the Bernstein-Bézier polynomials of degree n over triangles is presented. The condition is proved to be necessary also for $n = 2$ and $n = 3$.

Keywords: Bernstein-Bézier polynomials over triangles   Bézier ordinates   convexity

1 Introduction

Bernstein-Bézier triangular patches have recently and extensively studied in the context of CAGD[1,4,5,7,8].

Let T be a given triangle with vertices T_1, T_2, T_3. For point $p \in T$, let (u, v, w) be barycentric coordinates of p with respect to T. It is well-known that $0 \leqslant u, v, w \leqslant 1$ and $u + v + w = 1$. To each set $f = \{f_{i,j,k} : i, j, k \text{ are nonnegative integers such that } i + j + k = n\}$ the nth Bernstein-Bézier polynomial

$$B_n(f; p) = \sum_{i+j+k=n} f_{i,j,k} J_{i,j,k}^n(p) \tag{1}$$

is associated, in which

$$J_{i,j,k}^n(p) = \frac{n!}{i! j! k!} u^i v^j w^k, \quad i + j + k = n. \tag{2}$$

$f = \{f_{i,j,k}\}$ is called the set of Bézier ordinates of the polynomial (1). Geometrically (1) represents a triangular patch over T. For more detail we refer the readers to [4, 5].

Chang and Davis[2] initiate the investigations of convexity of the Bernstein-Bézier triangular patches. The function $F(p)$ is said to be convex over T if the following inequality

$$F\left(\frac{p_1+p_2}{2}\right) \leqslant \frac{1}{2}[F(p_1)+F(p_2)]$$

holds for any p_1, p_2 in T. According to [2], the set f is said to be convex in 3-direction, if the Bézier ordinates satisfy

$$\begin{aligned}\Delta^{(1)}_{i,j,k} &= (E_2-E_1)(E_3-E_1)f_{i,j,k} \geqslant 0,\\ \Delta^{(2)}_{i,j,k} &= (E_3-E_2)(E_1-E_2)f_{i,j,k} \geqslant 0,\\ \Delta^{(3)}_{i,j,k} &= (E_1-E_3)(E_2-E_3)f_{i,j,k} \geqslant 0,\end{aligned} \quad (3)$$

for all i, j, k, such that $i+j+k = n-2$, where E_1, E_2 and E_3 are characterized by

$$\left.\begin{aligned}E_1 f_{i,j,k} &:= f_{i+1,j,k}\\ E_2 f_{i,j,k} &:= f_{i,j+1,k}\\ E_3 f_{i,j,k} &:= f_{i,j,k+1}\end{aligned}\right\} \quad \text{for} \quad i+j+k = n-1.$$

The above algebraic definition of "convex in 3-direction" can be equivalently described geometrically as "each pair of adjacent subtriangles in the Bézier net forms a convex surface". For the definition of Bézier net, the readers may consult the papers by Farin, just mentioned above.

The authors of present paper showed in an unpublished paper (1983) that f is convex in 3-direction if and only if the corresponding Bézier net is convex over T.

One of the main theorems due to [2] is that if f is convex in 3-direction, then $B_n(f;p)$ is convex over T.

Condition (3) is sufficient rather than necessary for the convexity of $B_n(f;p)$ over T. Let us consider the following example: a Bernstein-Bézier triangular patch is defined by the following data: all the Bézier ordinates are zero except

$$f_{4,0,0} = f_{0,4,0} = f_{0,0,4} = 1 \quad \text{and} \quad f_{0,2,2} = f_{2,0,2} = f_{2,2,0} = \frac{1}{3}.$$

The corresponding Bernstein-Bézier polynomial is

$$B_4(f;p) = (u^2 + v^2 + w^2)^2. \tag{4}$$

Even though in this case f is not convex in 3-direction, one can verify without difficulty that $B_4(f;p)$ is convex over T.

In this paper another condition which is weaker than (3) for the convexity of $B_n(f;p)$ is provided. It will be shown that the condition is not only sufficient but also necessary for the convexity of $B_n(f;p)$ in the cases of $n=2$ and $n=3$.

2 Main results

Combination (12) and Theorem 3 both in [2] gives that $B_n(f;p)$ is convex over T iff the quadratic form

$$\begin{bmatrix} \xi, & \eta, & \zeta \end{bmatrix} \begin{bmatrix} \frac{\partial^2 B}{\partial u^2} & \frac{\partial^2 B}{\partial u \partial v} & \frac{\partial^2 B}{\partial u \partial w} \\ \frac{\partial^2 B}{\partial u \partial v} & \frac{\partial^2 B}{\partial v^2} & \frac{\partial^2 B}{\partial v \partial w} \\ \frac{\partial^2 B}{\partial u \partial w} & \frac{\partial^2 B}{\partial v \partial w} & \frac{\partial^2 B}{\partial w^2} \end{bmatrix} \begin{bmatrix} \xi \\ \eta \\ \zeta \end{bmatrix} \geqslant 0 \tag{5}$$

for all $p \in T$ and all (ξ, η, ζ) such that $\xi + \eta + \zeta = 0$. In (5), $B_n(f;p)$ has been represented by B for the sake of simplicity. Note that the matrix in (5) is the Hessian of B. It should be pointed out that (ξ, η, ζ) such that $\xi + \eta + \zeta = 0$ represents a direction in barycentric coordinates and that the left-hand side is just the second directional derivative of B with respect to the direction (see [4,5] again). Simple calculations yield

$$\frac{\partial^2 B}{\partial u^2} = n(n-1) \sum_{i+j+k=n-2} E_1^2 f_{i,j,k} J_{i,j,k}^{n-2}(p),$$

$$\frac{\partial^2 B}{\partial v \partial w} = n(n-1) \sum_{i+j+k=n-2} E_2 E_3 f_{i,j,k} J_{i,j,k}^{n-2}(p),$$

etc. (5) can be rewritten as

$$\sum_{i+j+k=n-2} Q_{i,j,k}(\xi,\eta,\zeta) J_{i,j,k}^{n-2}(p) \geq 0 \tag{6}$$

where the quadratic form $Q_{i,j,k}$ is defined by

$$Q_{ijk} = \begin{bmatrix} \xi, & \eta, & \zeta \end{bmatrix} \begin{bmatrix} E_1^2 f_{i,j,k} & E_1 E_2 f_{i,j,k} & E_1 E_3 f_{i,j,k} \\ E_1 E_2 f_{i,j,k} & E_2^2 f_{i,j,k} & E_2 E_3 f_{i,j,k} \\ E_1 E_3 f_{i,j,k} & E_2 E_3 f_{i,j,k} & E_3^2 f_{i,j,k} \end{bmatrix} \begin{bmatrix} \xi \\ \eta \\ \zeta \end{bmatrix}, \tag{7}$$

for $i+j+k=n-2$.

Let us begin with $n=2$. In this case (6) becomes

$$\begin{bmatrix} \xi, & \eta, & \zeta \end{bmatrix} \begin{bmatrix} A & c & b \\ c & B & a \\ b & a & C \end{bmatrix} \begin{bmatrix} \xi \\ \eta \\ \zeta \end{bmatrix} \geq 0, \tag{8}$$

where

$$A := f_{2,0,0}, \quad B := f_{0,2,0}, \quad C := f_{0,0,2}, \tag{9}$$
$$a := f_{0,1,1}, \quad b := f_{1,0,1}, \quad c := f_{1,1,0}.$$

It has been shown that (8) holds for $\xi+\eta+\zeta=0$ iff the 2×2 matrix

$$\begin{bmatrix} A+C-2b & C+c-a-b \\ C+c-a-b & B+C-2a \end{bmatrix} \tag{10}$$

is positive semidefinite. With

$$\Delta_a := A+a-b-c,$$
$$\Delta_b := B+b-c-a, \tag{11}$$
$$\Delta_c := C+c-a-b,$$

then the matrix in (10) reduces to

$$\begin{bmatrix} \Delta_a + \Delta_c & \Delta_c \\ \Delta_c & \Delta_b + \Delta_c \end{bmatrix}. \tag{12}$$

The matrix in (12) is positive semidefinite iff

$$\Delta_a + \Delta_c \geqslant 0, \quad \Delta_b + \Delta_c \geqslant 0,$$
$$(\Delta_a + \Delta_c)(\Delta_b + \Delta_c) - \Delta_c^2 \geqslant 0,$$

or, more symmetrically but equivalently,

$$\begin{aligned}\Delta_b + \Delta_c \geqslant 0, \quad \Delta_c + \Delta_a \geqslant 0, \quad \Delta_a + \Delta_b \geqslant 0, \\ \Delta_b \Delta_c + \Delta_c \Delta_a + \Delta_a \Delta_b \geqslant 0.\end{aligned} \quad (13)$$

Hence we have proven the following:

Theorem 1 $B_2(f;p)$ is convex over T iff (13) holds, where $\Delta_a, \Delta_b, \Delta_c$ are given by (11) with quantities defined in (9).

Theorem 2 $B_3(f;p)$ is convex over T iff

$$\begin{aligned}\Delta_{i,j,k}^{(2)} + \Delta_{i,j,k}^{(3)} \geqslant 0, \quad \Delta_{i,j,k}^{(3)} + \Delta_{i,j,k}^{(1)} \geqslant 0, \quad \Delta_{i,j,k}^{(1)} + \Delta_{i,j,k}^{(2)} \geqslant 0, \\ \Delta_{i,j,k}^{(2)} \Delta_{i,j,k}^{(3)} + \Delta_{i,j,k}^{(3)} \Delta_{i,j,k}^{(1)} + \Delta_{i,j,k}^{(1)} \Delta_{i,j,k}^{(2)} \geqslant 0,\end{aligned} \quad (14)$$

for $i + j + k = 1$.

Proof If $n = 3$, then (6) becomes

$$uQ_{1,0,0}(\xi,\eta,\zeta) + vQ_{0,1,0}(\xi,\eta,\zeta) + wQ_{0,0,1}(\xi,\eta,\zeta) \geqslant 0. \quad (15)$$

As has been shown before, condition (14) implies that

$$\begin{aligned}Q_{1,0,0}(\xi,\eta,\zeta) \geqslant 0, \\ Q_{0,1,0}(\xi,\eta,\zeta) \geqslant 0, \\ Q_{0,0,1}(\xi,\eta,\zeta) \geqslant 0,\end{aligned} \quad (16)$$

for all (ξ,η,ζ) such that $\xi + \eta + \zeta = 0$. Since the left-hand side of (15) is convex linear combination of $Q_{1,0,0}, Q_{0,1,0}, Q_{0,0,1}$, (15) holds for each $(u,v,w) \in T$ and $\xi + \eta + \zeta = 0$. Hence $B_3(f;p)$ is convex over T.

Conversely, we assume that $B_3(f;p)$ is convex over T, thus (15) holds for each $(u,v,w) \in T$. Taking in turn $u = 1$, $v = 1$ and $w = 1$, we obtain (16) from (15). Hence the condition (14) must hold. □

Generally, for each $p \in T$, the left-hand side of (6) is convex linear combination of $Q_{i,j,k}\,(\xi,\eta,\zeta)$ in which $i+j+k=n-2$ and $\xi+\eta+\zeta=0$. Now follows immediately

Theorem 3 If (14) holds for all $i+j+k=n-2$, then $B_n(f;p)$ is convex over T.

Since it is obvious that (3) implies (14), we get as a special case of Theorem 3 the following

Corollary 1 If f is convex in 3-direction, then $B_n(f;p)$ is convex over T.

Given

$$f_{2,0,0} = f_{0,2,0} = -f_{1,1,0} = 1,$$
$$f_{0,0,2} = f_{0,1,1} = f_{1,0,1} = 0,$$

the associated $B_2(f;p)$ is convex over T as it is justified by Theorem 1. But Chang and Davis' criterion fails to be valid as the set of Bézier ordinates is not convex in 3-direction. Hence we see that the present criterion is superior to that of Chang and Davis.

Note that for $n \geqslant 4$ the condition (14) is still a sufficient rather than a necessary condition for the convexity of $B_n(f;p)$. The example in the introductory part of this paper also works to provide a counter-example.

References

[1] Barnhill R E, Farin G. C^1 Quintic Interpolation over Triangles: Two Explicit Representations[J]. Int. J. Numer. Meth. Eng., 1981, 17: 1763-1778.

[2] Chang G, Davis P J. The Convexity of Bernstein Polynomials over Triangles[J]. J. Approx. Theory, 1984, 40: 11-28.

[3] Chang G, Feng Y-Y. A New Proof of the Convexity of the Bernstein Polynomials over Triangles[J]. Chinese Annals of Math., 1983.

[4] Farin G. Subsplines uber Dreiecken[R]. Braunschweig,1979.

[5] Farin G. Smooth Interpolation to Scattered 3D Data[J]. Computer Aided Geometric Design, 1983: 43-63.

[6] Goldman R N. Using Degenerate Bézier Triangles and Tetrahedra to Subdivide Bézier Curves[J]. Computer-aided Design, 1982,14: 307-311.

[7] Goldman R N. Subdivision Algorithms for Bézier Triangles[J]. Computer-aided Design, 1983, 15: 159-166.

[8] Sabin M A. The Use of Piecewise Forms for the Numerical Representation of Shape[R]. Budapest, 1977.

9

第一次国际会议

    1984 年，对我来说是一个丰收的年份，我在美国做访问学者的时候，我投稿的文章陆续刊登，更可喜的是，没有一篇被拒. 我与 Davis 的《三角域的凸性》，发表在美国著名《逼近论杂志》上 (40 卷 11~28 页)，是新年第一本杂志，我们的文章，非常靠前. 我是一个名不见经传的中年人，另一个是大名鼎鼎的应用数学家 (曾经当任过《逼近论杂志》编委)，著作等身.

    又是 1984 年，一个崭新的国际杂志《Computer Aided Geometric Design》(CAGD) 诞生了，由美国的 Barnhill 教授和西德 Boehm 教授担任主编，我被聘为中国的唯一编委. 我想，主要是因为 Barnhill 的推荐，而且他找不到其他的中国 CAGD 工作者. 这是一个很大的幸运，对我来说，好处很大. 首先，最新的杂志从荷兰的出版社用最短的路径寄到中国科大. 17 年来，CAGD 杂志，放在我们小组保管. 我卸下编委之后，也马上退休了，把所有的 CAGD 杂志捐赠给我们的小组，祝愿我们的小组，日新月异，发扬光大.

    两位主编组织的学术会议，也如近水楼台先得月，编委们经常受到邀请.

    1984 年 11 月上旬，在联邦德国森林里的 Oberwolfach 数学研究所举行了 "计算机辅助设计中的曲面方法" 会议，由 Barnhill 和 Boehm 主持. 在数学所免费住宿，感谢 Oberwolfach 数学所，他们慷慨地为我提供往返的国际机票.

    中国俗语讲到 "福无双至"，也不尽然. 1984 年上半年，我收到来自联邦德国的一封信，寄信人对我来讲完全陌生. 打开一看，是 Siegen 大学数学系的 Walter Schempp 寄来的，邀请我参加 1985 年 1 月 20 日在 Oberwolfach 数学所召开的"多维构造性逼近论"会议，会议由 Walter Schempp 本人主持.

他的信是专门为我写的. 他提到今年初发表在《逼近论杂志》的凸性文章，称赞我是一名"专家"，希望与我能在 Oberwolfach 数学所相见.

两个会议之间相差两个月，回中国去然后来西德，我估计德国的数学所不可能再掏国际机票. 这时候，我想起我们 CAGD 编委会的德国同仁，虽然我们还没有见过面. 一位是 Darmstadt 工业大学数学系教授 J. Hoschek(1935~2009)，另一位是 Berlin 工业大学造船系教授 H. Nowacki，他们都同意为我申请经费在他们的大学里访问一个月. 德国人讲话都是算数的，我们都成了很好的朋友. 德国人讲英语有自己的口音，但是 Nowacki 从小在美国待过，美语非常纯正. 他是搞造船的，与数学有一些距离. Hoschek 是搞几何的，在德国是顶级的教授. 冯玉瑜教授访问过 Hoschek 并联合发表过文章.

几经周折，到西德开两次会议，已经落实. 那时候没有 Email，没有国际直拨到中国的电话，全凭航空邮件，万一信件丢失，非常误事. 总之，我一切幸运.

1984 年 11 月上旬的一个晴朗的日子，我在法兰克福走下飞机，第一次踏上了联邦德国的土地. 换上了南下的火车，电气火车以每个小时百余公里的速度奔驰，却又是那么平稳，并不使人明显感到钢轨接头处产生的振动. 整修得很好的田野，大片的绿色草地，错落有致的村舍，向后飞去，使人感到清洁而宁静，充满生机. 我在名叫"郝萨赫"的小站下车，出租车把我带到了半山之中，在数学所的大门口停了下来.

实际上，这是一座小山的山顶，抬头望去，四周群山环绕，山上层层覆盖着针叶红杉. 俯瞰山下，只见沿山势而盘绕的公路上车辆来往不断，形状各不相同的房屋散布在公路两旁. 这里空气新鲜，恬静，到了深夜，甚至可以听到山脚下小溪流水的声响.

我到 Oberwolfach 数学所来参加国际会议，一共有三次：1984 年，1985 年和 1987 年. 每一次都被其浓厚的学术氛围、朴实无华的舒适生活所吸引，我曾经在科大校刊上发表连载文章，来宣扬 Oberwolfach. 后来，30 多年过去了，我再也没有到过 Oberwolfach，情况一定有很大的变化. 如今，来 Oberwolfach 开会的中国同行比比皆是，司空见惯.

现在，我把 Oberwolfach 在 80 年代的情况简单描述一下. 当时只有两座建筑，一座叫做"宾馆"，另一座叫"会议楼". 宾馆的第一层有厨房、餐厅和酒吧，还有数学所的几间办公室. 这儿的电话间可以向世界上发达国家和地区直接拨号，但当时不能打往中国. 第二层和第四层有客房和卫生设备，客房里陈设简朴而适用. 就是在任何客房里都没有电视，大约这是为了避免干扰，让人们能专心致志于学术讨论.

会议楼的第一层，主要是一个相当完善的图书馆，当代世界数学期刊，应有尽有. 数学书籍馆藏丰富，供人们自由查阅，但不能带出楼外. 第二层是报告厅，这里装有六块可以上下移动的黑板，投影仪、幻灯机乃至电影放映机可供随时使用.

在这个数学所召开的专业会议，最长不得超过五天，与会者凭邀请信在星期天下午来所报到，星期六早餐后必须离开数学所，以便服务人员对客房、餐厅、报告厅作彻底的清扫，准备迎接下一个会议.

在正式报告之外，与会者还有充分的时间和机会私下接触，在饭桌上，在酒吧里，或促膝谈心，或推演公式. 每个星期三下午，不安排大小报告，代之的是一次远足. 人们三五成群，边走边谈，也观看山景，按照预定的路线，到某一个小酒店汇合，稍事休息，饮上一杯咖啡或茶，吃上一块点心 (自己付费)，再沿着新的道路返回数学所. 主人的精心安排令我叹服. 第一次就餐时，每一个人的餐巾，放在一个塑料套子里，套子的外面印有各人的名字和编号. 就餐之前，你得先找到自己的餐巾，对号入座. 每餐开饭之前，餐巾由服务员随机组合，为的是让每一个人都能同尽可能多的人相遇.

这是我第一次参加 CAGD 国际会议，除了老朋友 Barnhill 和 Farin，大部分 CAGD 的知名学者我都是第一次见到. 我结识了 Thomas W. Sederberg，他是美国 Brigham Young 大学土木工程系的教授. 那时候他非常年轻，他带着他的妻子 Branda Sederberg，简直是郎才女貌，一对金童玉女. 我记得他所做报告的题目是"平面分片代数曲线". 我与 Sederberg 的故事，以后加以叙述.

大会刚一结束，我就告别 Oberwolfach 数学所，这是一个星期五，我坐着 Hoschek 的汽车，往 Darmstadt 方向驶去，他的大学就在那里，我在他的数学系访问一个月.

我准备与 Hoschek 共同发表一篇文章，时间还是很紧的，天天待在我的办公室里. 有一次 Hoschek 在他的家里宴请我，我送他的夫人一条真丝头巾.

后来，他们夫妇俩开车到 Heideberg 陪我旅游，记得有一个大酒桶奇大无比，据说有一个酒鬼被淹死在酒桶里. 我们三个买了三杯酒，一饮而尽. 我们的酒杯可以带走，作为纪念. 他们夫妇的酒杯都送给了我，三个酒杯至今仍在我合肥的家中.

一个月以后，我到 (西) 柏林工业大学访问，主人是 Nowacki. 那个时候，德国分成东西两个国家，从 Darmstadt 到东柏林，再到西柏林，两次过关，非常不方便. Nowacki 对中国学者极为友善，这一个月的访问，他没有对我提任何要求，我没有丝毫压力. 冯玉瑜和我带的第一批硕士研究生，其中之一就是闫新民，毕业之后分到南京航空学院，后来到柏林工业大学读 Nowacki 的博士，1994 年博士毕业.

在柏林期间，有两位校友给予我很多的帮助：黄少光、刘晋夫妇，他们是同班同学，在中国科大，我还不认得他们. 他们带我参观博物馆，游览东柏林，在最高的旋转餐厅吃过饭.

1985 年 1 月 20 日，我来 Oberwolfach 参加第二个国际会议. 这一次是从 Stuttgart 乘上火车，走与第一次不同的路线. 时间才过去两个多月，但自然景色已经迥异：大地被白雪覆盖，黑森林已经变成"白森林"了.

这一次会议主题是"多元逼近论"，由 Siegen 大学数学系 Schempp 教授等主持. 我在这一会议上也做了报告，由于两次会议参加人员有交集，我两个报告内容不同. 这次来了三位中国学者：北京大学的沈燮昌教授、吉林大学的王仁宏教授和我. 我和沈教授不认识，但知其大名. 在那个时候，一次会议来了三位中国学者，非常稀罕.

下面的文章原载于：《Journal of Approximation Theory》，1989 年，58 卷第 247~258 页.

Convergence of Bézier Triangular Nets and a Theorem of Pólya

G. Z. Chang[①]   J. Hoschek

Department of Mathematics, Technische Hochschule Darmstadt,
6100 Darmstadt, West Germany
Communicated by Charles K. Chui
Received June 16, 1987

Abstract: This paper is concerned with Bernstein-Bézier triangular patches and their Bézier nets. By degree raising, a sequence of Bézier nets is obtained. It is known that the sequence converges uniformly to the Bernstein-Bézier triangular patch determined by those nets. A new proof of the convergence, which is more geometric and constructive, is presented. Connections of convergence and a theorem due to Pólya are revealed. Extensions to higher dimensional cased are also mentioned.

1 Introduction

Let T be a given triangle. Each point P in T has barycentric coordinates (u,v,w) with respect to T. The triple (u,v,w) satisfies the conditions

$$u \geqslant 0, \quad v \geqslant 0, \quad w \geqslant 0,$$
$$u+v+w = 1. \tag{1}$$

We identify P and its barycentric coordinates by writing $P = (u,v,w)$. Let n be any positive integer.

[①] Department of Mathematics, University of Science and Technology of China, Hefei, Anhui, P. R. China.

The subdivision of T into n^2 congruent triangles with vertices at $(\frac{i}{n}, \frac{j}{n}, \frac{k}{n})$, in which $i+j+k=n$, denoted by $S_n(T)$, is called the nth subdivision of T. The points $(\frac{i}{n}, \frac{j}{n}, \frac{k}{n}), i+j+k=n$, are called nodes of $S_n(T)$. $S_4(T)$ is illustrated in Fig 1.

Given is a set f of $\frac{(n+1)(n+2)}{2}$ real numbers, i.e., $f := \{f_{i,j,k} | i+j+k=n\}$, and the polynomial

$$B^n(f;p) := \sum_{i+j+k=n} f_{i,j,k} \frac{n!}{i!j!k!} u^i v^j w^k \qquad (2)$$

is defined as the Bernstein-Bézier(B-B) polynomial of f over the triangle T.

Fig 1 $S_4(T)$ with its nodes

$f_{i,j,k}(i+j+k=n)$ are called the Bézier ordinates of $B^n(f;p)$ while $(\frac{i}{n}, \frac{j}{n}, \frac{k}{n}; f_{i,j,k})$ are called its Bézier points. The point set $(P; B^n(f;p))$ with $p \in T$ forms a surface patch over triangle T. We simply call polynomial (2) the B-B triangular patch over triangle T. The piecewise linear function $\hat{f}(p)$ which is linear on each subtriangle of $S_n(T)$ and interpolates to $f_{i,j,k}$ at $(\frac{i}{n}, \frac{j}{n}, \frac{k}{n})$, is said to be the Bézier net of patch (2). Figure 2 illustrates a Bézier net and the corresponding patch (with $n=3$).

It is known[3] that if we set

$$E f_{i,j,k} := \frac{1}{n+1}(i f_{i-1,j,k} + j f_{i,j-1,k} + k f_{i,j,k-1}), \qquad (3)$$

where $i+j+k=n+1$, and write

$$Ef := \{Ef_{i,j,k} | i+j+k = n\},$$

then we have

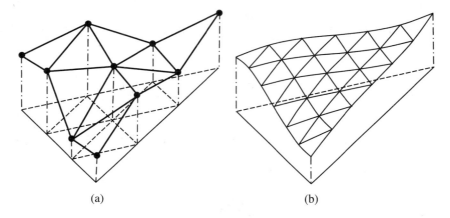

(a)　　　　　　　　　　　(b)

Fig 2　Bézier net and the corresponding patch (n=3)

$$B^n(f;p) = B^{n+1}(Ef;p).$$

This means that it is always possible to write $B^n(f;p)$ as a B-B polynomial of degree $n+1$. The technique just mentioned is called degree raising. The Bézier net associated with Ef is denoted by $E\hat{f}(p)$ which is linear on each subtriangle of $S_{n+1}(T)$ and interpolates $Ef_{i,j,k}$ at $(\dfrac{i}{n+1}, \dfrac{j}{n+1}, \dfrac{k}{n+1})$.

If one repeats the process of degree raising, a sequence of Bézier nets $\hat{f}(p), E\hat{f}(p), E^2\hat{f}(p), \cdots$, will be obtained. It has been proved that

Theorem 1　*We have*

$$\lim_{m \to \infty} E^m \hat{f}(p) = B^n(f;p) \tag{4}$$

uniformly on T.

Recently we found that Theorem 1 has a very close connection with a famous theorem, which appeared in the early stage of this century[6], in the algebraic theory of polynomials in several variables. For historical remarks, see [5]. To present Pólya's theorem we need some definitions. A real *form* is a homogeneous polynomial $F(x_1, x_2, \cdots, x_m)$, with real coefficients, in m variables. A form is said to be strictly positive, in a certain region of the variables, if $F > 0$ for all points in that region.

Theorem 2 [Pólya] *If the form $F(x_1, x_2, \cdots, x_m)$ is strictly positive in the region*

$$(x_1, x_2, \cdots, x_m), \quad x_1 \geqslant 0, x_2 \geqslant 0, \cdots, x_m \geqslant 0$$

and

$$x_1 + x_2 + \cdots + x_m > 0,$$

then F may be expressed as

$$F = \frac{G}{H}, \tag{5}$$

where G and H are forms with positive coefficients. In particular, we may suppose that

$$H = (x_1 + x_2 + \cdots + x_m)^p$$

for a suitable natural number p.

In the present paper, we first show that Theorem 2 can be derived from Theorem 1, and then point out that Pólya's technique for the proof of his theorem, with further modifications, in turn provides a proof for Theorem 1 which is more geometric and constructive than existing ones.

2 Proof of Theorem

For simplicity of writing we suppose $m = 3$. No new point of principle arises for general m.

A form F in three variables u, v, w can be expressed by

$$F(u, v, w) = \sum_{\alpha+\beta+\gamma=n} f_{\alpha,\beta,\gamma} \frac{n!}{\alpha!\beta!\gamma!} u^\alpha v^\beta w^\gamma, \tag{6}$$

in which u, v, w are independent. If $F > 0$ in the region $u \geqslant 0, v \geqslant 0, w \geqslant 0$ and $u + v + w > 0$, then F has a positive minimum, say τ, in the region $u \geqslant 0, v \geqslant$

$0, w \geqslant 0$ and $u+v+w=1$. In this case (6) becomes a B-B polynomial on the triangle T.

An elementary manipulation brings the following identity
$$(u+v+w)^m F = \frac{m!n!}{(m+n)!} \sum_{a+b+c=m+n} \sum_{i+j+k=n} f_{i,j,k} \times \binom{a}{i}\binom{b}{j}\binom{c}{k} \frac{(a+b+c)!}{a!b!c!} u^a v^b w^c. \tag{7}$$

For a proof, see [5, 8]. If $(u,v,w) \in T$, i.e., $u+v+w=1$, then (7) can be viewed as the mth degree raising of the B-B polynomial $B^n(f;p)$. Hence we have
$$E^m f_{i,j,k} = \frac{m!n!}{(m+n)!} \sum_{\alpha+\beta+\gamma=n} f_{\alpha,\beta,\gamma} \binom{i}{\alpha}\binom{j}{\beta}\binom{k}{\gamma}, \tag{8}$$
in which $i+j+k=m+n$.

By Farin's theorem, the inequality
$$E^m \hat{f}(p) \geqslant \frac{\tau}{2} > 0$$
holds for all P in T and for sufficiently large m. Particularly,
$$E^m f_{i,j,k} = E^m \hat{f}\left(\frac{i}{m+n}, \frac{j}{m+n}, \frac{k}{m+n}\right) > 0 \tag{9}$$
for $i+j+k=m+n$. We denote the form in the right-hand side of (7) by G_m. Equation (9) shows that all coefficients of G_m are positive for sufficiently large m. Identity (7) gives
$$F(u,v,w) = \frac{G_m(u,v,w)}{(u+v+w)^m}$$
which is the desired representation for sufficiently large m.

The strict positivity of B-B polynomials was characterized by Zhou[8]. Obviously he was not aware of Theorem 2.

3 An alternate proof of Theorem 1

There are several proofs for the theorem. The original proof[3] is very short but some sophisticated results by Stancu are involved. The proof given by Zhou ([8]; see also [4]) is relatively elementary but it does not provide a proof for the *uniform* convergence. For other proofs the reader is referred to [1,7] in which the structure of Bézier nets has been carefully studied.

We define for real x and nonnegative i the usual binomial coefficient $\binom{x}{i}$ as

$$\binom{x}{0} = 1,$$

$$\binom{x}{i} = \frac{x(x-1)\cdots(x-i+1)}{i!}, \quad i = 1, 2, 3, \cdots.$$

Consider the following polynomial of degree n:

$$L_n(f;p) = \sum_{\alpha+\beta+\gamma=n} f_{\alpha,\beta,\gamma} \binom{nu}{\alpha}\binom{nv}{\beta}\binom{nw}{\gamma}. \tag{10}$$

It is easy to verify that

$$L_n\left(f; \frac{i}{n}, \frac{j}{n}, \frac{k}{n}\right) = f_{i,j,k}, \quad i+j+k = n.$$

This means that (10) is the Lagrange interpolation to $\hat{f}(p)$ at all nodes of $S_n(T)$.

In particular, if $f_{\alpha,\beta,\gamma} = 1$ for all $\alpha+\beta+\gamma = n$, from (10) we have the following identity,

$$\sum_{\alpha+\beta+\gamma=n} \binom{nu}{\alpha}\binom{nv}{\beta}\binom{nw}{\gamma} = 1,$$

for $u+v+w = 1$ and $n = 1, 2, 3, \cdots$. In general, we have

$$\sum_{\alpha+\beta+\gamma=n} \binom{a}{\alpha}\binom{b}{\beta}\binom{c}{\gamma} = \binom{a+b+c}{n}. \tag{11}$$

The Lagrange interpolation to $E^m \hat{f}(p)$ at all nodes of $S_{m+n}(T)$, by (10) and (8), is

$$\frac{m!n!}{(m+n)!} \sum_{i+j+k=n} f_{i,j,k} \sum_{\alpha+\beta+\gamma=m+n} \binom{\alpha}{i}\binom{\beta}{j}\binom{\gamma}{k}\binom{A}{\alpha}\binom{B}{\beta}\binom{C}{\gamma}, \tag{12}$$

in which $A := (m+n)u, B := (m+n)v, C := (m+n)w$. It is clear that

$$\binom{\alpha}{i}\binom{A}{\alpha} = \binom{A}{i}\binom{A-i}{\alpha-i},$$

$$\binom{\beta}{j}\binom{B}{\beta} = \binom{B}{j}\binom{B-j}{\beta-j},$$

$$\binom{\gamma}{k}\binom{C}{\gamma} = \binom{C}{k}\binom{C-k}{\gamma-k},$$

and by (11) that

$$\sum_{\alpha+\beta+\gamma=m+n} \binom{A-i}{\alpha-i}\binom{B-j}{\beta-j}\binom{C-k}{\gamma-k} = 1,$$

as

$$A-i+B-j+C-k = (A+B+C)-(i+j+k) = m+n-n = m$$

and

$$\alpha-i+\beta-j+\gamma-k = m+n-n = m.$$

Hence (12) becomes

$$\frac{m!n!}{(m+n)!}\sum_{i+j+k=n} f_{i,j,k} \binom{(m+n)u}{i}\binom{(m+n)v}{j}\binom{(m+n)w}{k}. \quad (13)$$

Define

$$\varphi(u,v,w;t) := n!t^n \sum_{i+j+k=n} f_{i,j,k} \binom{ut^{-1}}{i}\binom{vt^{-1}}{j}\binom{wt^{-1}}{k}, \quad 0 \leqslant t \leqslant 1. \quad (14)$$

We can verify that (13) is equal to

$$\lambda_m \varphi\left(u,v,w;\frac{1}{m+n}\right), \quad (15)$$

where

$$\lambda_m := \frac{(m+n)^n}{(m+1)(m+2)\cdots(m+n),} \quad (16)$$

$m = 1, 2, 3, \cdots$. It is obvious that $\lim_{t \to 0} \varphi(p; t) = F(p)$, as

$$t^i \begin{pmatrix} ut^{-1} \\ i \end{pmatrix} = \frac{u(u-t)(u-2t)\cdots[u-(i-1)t]}{i!} \to \frac{u^i}{i!} \quad (t \to 0),$$

etc. If we define $\varphi(p; 0) := F(p)$, then $\varphi(p; t)$ is continuous on the region

$$u \geqslant 0, \quad v \geqslant 0, \quad w \geqslant 0, \quad u + v + w = 1, \quad 0 \leqslant t \leqslant 1. \tag{17}$$

Function $\varphi(p; t)$ represents a family of surfaces with a single parameter $t \in [0, 1]$. Especially we have mentioned that the surface patch $\lambda_m \varphi(p; \frac{1}{m+n})$ coincides with the Bézier net $E^m \hat{f}(p)$ at all its vertices.

The investigation of convergence of $E^m \hat{f}(p)$ is now shifted to that of $\lambda_m \varphi(p; \frac{1}{m+n})$. The second problem is easier than the first as $\varphi(p; t)$ has an analytical expression on T, while $E^m \hat{f}(p)$, being a piecewise linear function, does not. By the mean value theorem of univariate functions we know that

$$\varphi(p; t') - \varphi(p; t) = O(|t' - t|), \quad t' \to t$$

in the region (17). In particular we have

$$\varphi\left(p; \frac{1}{m+n}\right) - F(p) = \varphi\left(p; \frac{1}{m+n}\right) - \varphi(p; 0) = O\left(\frac{1}{m}\right), \quad \text{as} \quad m \to \infty.$$

Since $\lambda_m = 1 + O(\frac{1}{m})$, we still have

$$\lambda_m \varphi\left(p; \frac{1}{m+n}\right) - F(p) = O\left(\frac{1}{m}\right). \tag{18}$$

We have shown that the sequence of surfaces $\lambda_m \varphi(p; \frac{1}{m+n})$ converges uniformly to the B-B patch $B^n(f; p)$ with the rate $O(\frac{1}{m})$ as $m \to \infty$. Now we have to estimate the difference between $\lambda_m \varphi(p; \frac{1}{m+n})$ and the corresponding Bézier net $E^m \hat{f}(p)$.

Take a typical upward subtriangle with vertices

$$\left(\frac{i+1}{m+n}, \frac{j}{m+n}, \frac{k}{m+n}\right),$$

$$\left(\frac{i}{m+n}, \frac{j+1}{m+n}, \frac{k}{m+n}\right),$$
$$\left(\frac{i}{m+n}, \frac{j}{m+n}, \frac{k+1}{m+n}\right),$$

in which $i+j+k = m+n-1$ (see Fig 3).

Let P be any point inside the subtriangle and P has the barycentric coordinates (λ, μ, ν) with respect to the subtriangle. Hence the barycentric coordinates (u, v, w) with respect to the domain triangle T will be

$$\left(\frac{i+\lambda}{m+n}, \frac{j+\mu}{m+n}, \frac{k+\nu}{m+n}\right).$$

Being linear on the subtriangle, $E^m \hat{f}(p)$ is a linear convex combination of its values at three vertices of the subtriangle; more precisely,

$$E^m \hat{f}(p) = \lambda E^m f_{i+1,j,k} + \mu E^m f_{i,j+1,k} + \nu E^m f_{i,j,k+1}$$

which becomes by (8)

$$E^m \hat{f}(p) = \frac{n!m!}{(n+m)!} \sum_{\alpha+\beta+\gamma=n} f_{\alpha,\beta,\gamma} \left[\lambda \binom{i+1}{\alpha} \binom{j}{\beta} \binom{k}{\gamma} \right.$$
$$\left. + \mu \binom{i}{\alpha} \binom{j+1}{\beta} \binom{k}{\gamma} + \nu \binom{i}{\alpha} \binom{j}{\beta} \binom{k+1}{\gamma} \right]. \tag{19}$$

On the other hand, by (14) we have

$$\lambda_m \varphi\left(p; \frac{1}{m+n}\right) = \lambda_m \varphi\left(\frac{i+\lambda}{m+n}, \frac{j+\mu}{m+n}, \frac{k+\nu}{m+n}; \frac{1}{m+n}\right)$$
$$= \frac{m!n!}{(m+n)!} \sum_{\alpha+\beta+\gamma=n} f_{\alpha,\beta,\gamma} \binom{i+\lambda}{\alpha} \binom{j+\mu}{\beta} \binom{k+\nu}{\gamma}. \tag{20}$$

Define

$$\psi(\lambda, \mu, \nu) := \binom{i+\lambda}{\alpha} \binom{j+\mu}{\beta} \binom{k+\nu}{\gamma}.$$

By Taylor expansion we obtain

$$\psi(\lambda, \mu, \nu) - \psi(1, 0, 0) = (\lambda - 1) \frac{\partial \psi}{\partial \lambda}(\lambda^*, \mu^*, \nu^*)$$
$$+ \mu \frac{\partial \psi}{\partial \mu}(\lambda^*, \mu^*, \nu^*) + \nu \frac{\partial \psi}{\partial \nu}(\lambda^*, \mu^*, \nu^*),$$

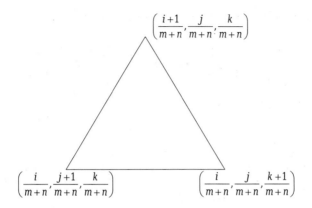

Fig 3　A typical upward subtriangle

where $(\lambda^*, \mu^*, \nu^*)$ is some point in T. It is clear that

$$\left|\binom{i+\lambda}{\alpha}\right| \leqslant \binom{i+\alpha}{\alpha},$$

$$\left|\binom{j+\mu}{\beta}\right| \leqslant \binom{j+\beta}{\beta},$$

$$\left|\binom{k+\nu}{\gamma}\right| \leqslant \binom{k+\gamma}{\gamma}.$$

Since

$$\frac{\partial}{\partial \lambda}\binom{i+\lambda}{\alpha} = \frac{1}{\alpha!}\sum_{\tau=0}^{\alpha-1}\frac{(\lambda+i)(\lambda+i-1)\cdots(\lambda+i-\alpha+1)}{\lambda+i-\tau},$$

similar estimation shows that

$$\left|\frac{\partial}{\partial \lambda}\binom{i+\lambda}{\alpha}\right| \leqslant \binom{\alpha+i}{\alpha-1}$$

and then

$$\left|\frac{\partial \psi}{\partial \lambda}\right| = \left|\frac{\partial}{\partial \lambda}\binom{i+\alpha}{\alpha}\right|\left|\binom{j+\mu}{\beta}\right|\left|\binom{k+\nu}{\gamma}\right|$$

$$\leqslant \binom{i+\alpha}{\alpha-1}\binom{j+\beta}{\beta}\binom{k+\gamma}{\gamma}.$$

Similar inequalities holds for $|\partial\psi/\partial\mu|$ and $|\partial\psi/\partial\nu|$. Therefore

$$\left|\binom{i+\lambda}{\alpha}\binom{j+\mu}{\beta}\binom{k+\nu}{\gamma} - \binom{i+1}{\alpha}\binom{j}{\beta}\binom{k}{\gamma}\right|$$
$$= |\psi(\lambda,\mu,\nu) - \psi(1,0,0)|$$
$$\leqslant \left|\frac{\partial\psi}{\partial\lambda}\right| + \left|\frac{\partial\psi}{\partial\mu}\right| + \left|\frac{\partial\psi}{\partial\nu}\right|$$
$$\leqslant \binom{i+\alpha}{\alpha-1}\binom{j+\beta}{\beta}\binom{k+\gamma}{\gamma} + \binom{i+\alpha}{\alpha}\binom{j+\beta}{\beta-1}\binom{k+\gamma}{\gamma}$$
$$+ \binom{i+\alpha}{\alpha}\binom{j+\beta}{\beta}\binom{k+\gamma}{\gamma-1}.$$

Hence by (10)

$$\left|\sum_{\alpha+\beta+\gamma=n} f_{\alpha,\beta,\gamma}\left[\binom{i+\lambda}{\alpha}\binom{j+\mu}{\beta}\binom{k+\nu}{\gamma} - \binom{i+1}{\alpha}\binom{j}{\beta}\binom{k}{\gamma}\right]\right|$$
$$\leqslant 3\|f\|\binom{m+2n-1}{n-1}$$

in which

$$\|f\| := \max\{|f_{\alpha,\beta,\gamma}| : \alpha+\beta+\gamma=n\}$$

and (11) has been used. Finally

$$\left|\lambda_m\varphi\left(p;\frac{1}{m+n}\right) - E^m\hat{f}(p)\right| \leqslant \frac{m!n!}{(m+n)!}3\|f\|\binom{m+2n-1}{n-1} = O\left(\frac{1}{m}\right). \tag{21}$$

The same estimate is valid for P lying on the downward subtriangles of $S_{m+n}(T)$. Combining (18) and (21) we get

$$F(p) - E^m\hat{f}(p) = O\left(\frac{1}{m}\right), \quad m \to \infty. \tag{22}$$

This completes the proof of Theorem 1.

4 Higher dimension cases

For $P \in \mathbb{R}^s$ and any $s+1$ affinely independent points $T_i \in \mathbb{R}^s$, $i=0,1,\cdots,s$, there are $s+1$ real numbers $\lambda_0, \lambda_1, \cdots, \lambda_s$ uniquely determined by de Boor[1]

$$P = \sum_{i=0}^{s} \lambda_i T_i$$

and

$$\sum_{i=0}^{s} \lambda_i = 1.$$

$(\lambda_0, \lambda_1, \cdots, \lambda_s)$ are called the barycentric coordinates of P with respect to the s-simplex T spanned by T_i, $i=0,1,\cdots,s$. We write $\lambda = (\lambda_0, \lambda_1, \cdots, \lambda_s)$. For a set i of $s+1$ nonnegative integers $i_0, i_1, \cdots, i_s$, we define

$$i := (i_0, i_1, \cdots, i_s),$$
$$|i| := i_0 + i_1 + \cdots + i_s,$$
$$i! = i_0! i_1! \cdots i_s!,$$
$$\lambda^i := \lambda_0^{i_0} \lambda_1^{i_1} \cdots \lambda_s^{i_s}.$$

With any set of scalars $f := \{f_i | \ |i| = n\}$, we define the Bernstein-Bézier polynomial on the simplex T by

$$B^n(f;\lambda) := \sum_{|i|=n} f_i \frac{n!}{i!} \lambda^i, \qquad (23)$$

in which f is called the set of Bézier ordinates for $B^n(f;\lambda)$. The degree raising technique is the same. It is easy to show that

$$E^m f_j = \frac{m! n!}{(m+n)!} \sum_{|i|=n} f_i \binom{j}{i}, \qquad (24)$$

where

$$\binom{j}{i} := \binom{j_0}{i_0} \binom{j_1}{i_1} \cdots \binom{j_s}{i_s},$$

and $|j| = m+n$. Formula (24) generalizes (8). For $s > 2$, it has been rightfully stressed and detailed by Dahmen and Micchelli[2] that, since there are several equally reasonable subdivisions of the simplex T, the Bézier nets $\hat{f}(\lambda)$ could not be uniquely determined. Similar to (14), we define

$$\varphi(\lambda;t) := n!t^n \sum_{|i|=n} f_i \begin{pmatrix} \lambda t^{-1} \\ i \end{pmatrix} \qquad (25)$$

for $t > 0$, where

$$\begin{pmatrix} \lambda t^{-1} \\ i \end{pmatrix} := \begin{pmatrix} \lambda_0 t^{-1} \\ i_0 \end{pmatrix} \begin{pmatrix} \lambda_1 t^{-1} \\ i_1 \end{pmatrix} \cdots \begin{pmatrix} \lambda_s t^{-1} \\ i_s \end{pmatrix}.$$

We can show easily that the function

$$\frac{(m+n)^n}{(m+1)(m+2)\cdots(m+n)} \varphi\left(\lambda; \frac{1}{m+n}\right)$$

interpolates to $E^m f_i$ at $\dfrac{i}{m+n}$ in which $|i|=m+n$ and that

$$\frac{(m+n)^n}{(m+1)(m+2)\cdots(m+n)} \varphi\left(\lambda; \frac{1}{m+n}\right) - B^n(f;\lambda) = O\left(\frac{1}{m}\right).$$

In a word, the results in previous sections of our paper can be extended in an obvious manner. Therefore we arrive at the extension of (20),

$$B^n(f;\lambda) - E^m \hat{f}(\lambda) = O\left(\frac{1}{m}\right),$$

in which $E^m \hat{f}(\lambda)$ denotes any reasonable piecewise linear interpolant to the data points

$$\left(\frac{i}{m+n}; E^m f_i\right), \quad |i| = n+m.$$

References

[1]   De Boor C. B-form Basics[R]. MRC Technical Summary Report No. 2957, 1986.

[2]   Dahmen W, Micchelli C. Convexity of Multivariate Bernstein Polynomials and Box Spline Surfaces[M]. Preprint No.735, Universitat Bonn, 1985.

[3]   Farin G. Ubsplines Über Dreiecken[R]. Braunschweig, FRG, 1979.

[4] Farin G. Triangular Bernstein-Bézier Patches[J]. J. CAGD, 1986, 3:83-127.

[5] Hardy G H, Littlewood J E, Pólya G. Nequalities[M]. London: Cambridge Univ. Press, 1952.

[6] Pólya G. Über Positive Darstellung von Polynomen[J]. Vierteljahresschr. Naturforsch. Ges. Zuerich, 1928, 73:141-145.

[7] Zhao K, Sun J. Dual Bases of Multivariate B-B Polynomials, preprint, submitted for publication.

[8] Zhou J. The Positivity and Convexity for Bernstein Polynomials over Triangles[J]. Math. Numer. Sinica, 1986, 8: 185-190.

Kelisky-Rivlin 定理

1967 年，数学杂志 Pacific J. of Math，21：511~520 上，有一篇文章叫做《Iterates of Bernstein Polynomials》，作者是 Kelisky 和 Rivlin，其中有一个定理是说：一个函数的 Bernstein 多项式经过反复迭代之后，其极限是一段直线. 这个定理非常简单明了和直观. 他们的证明用到了特征值和特征向量. 后来，我想到了更简单的证法，中学生都可以看得懂. 这个证法在我的小册子——《变换与数学竞赛》（中国少年儿童出版社，1993 年）中，就有记载. 好在篇幅很短，照抄如下：

设 f 在 $[0,1]$ 有定义，它的 n 次 Bernstein 多项式定义为

$$B_n(f;x) = \sum_{i=0}^{n} f\left(\frac{i}{n}\right) x^i (1-x)^{n-i},$$

简称 $B_n(f)$. 计算表明

$$\begin{aligned} B_n(1;x) &= 1, \\ B_n(x;x) &= x, \\ B_n(x^2;x) &= \frac{1}{n}x + \left(1-\frac{1}{n}\right)x^2, \end{aligned}$$

所以 $B_n(x(1-x);x) = \left(1-\dfrac{1}{n}\right)x(1-x).$

令 $\varphi(x) = f(x) - (1-x)f(0) - xf(1)$,得到
$$\varphi(0) = \varphi(1) = 0.$$

所以
$$B_n(\varphi) = B_n(f) - (1-x)f(0) - xf(1)$$
$$= \sum_{i=1}^{n-1} f\left(\frac{i}{n}\right)\binom{n}{i} x^i(1-x)^{n-i} + (1-x)^n f(0)$$
$$- (1-x)f(0) + x^n f(1) - xf(1).$$

不难证明存在常数 K_1, K_2, K_3 使得
$$\left|\sum_{i=1}^{n-1} f\left(\frac{i}{n}\right)\binom{n}{i} x^i(1-x)^{n-i}\right| \leqslant K_1 x(1-x),$$
$$|[(1-x)^n - (1-x)]f(0)| \leqslant K_2 x(1-x),$$
$$|(x^n - x)f(1)| \leqslant K_3 x(1-x).$$

置 $K = K_1 + K_2 + K_3$,使得
$$|B_n(\varphi)| \leqslant Kx(1-x), \quad x \in [0,1].$$

二次迭代,得
$$\left|B_n^{(2)}(\varphi)\right| \leqslant KB_n(x(1-x)) = K\left(1 - \frac{1}{n}\right)x(1-x),$$

三次迭代,就是
$$\left|B_n^{(3)}(\varphi)\right| \leqslant K\left(1 - \frac{1}{n}\right)^2 x(1-x),$$
$$\cdots$$

迭代趋向无穷,就是
$$\lim_{k \to \infty} B_n^{(k)}(\varphi; x) = 0.$$

最后得到
$$\lim_{k \to \infty} B_n^{(k)}(f; x) = (1-x)f(0) + xf(1),$$

这就是 Kelisky-Rivlin 定理.

后来我和冯玉瑜把 Kelisky-Rivlin 定理推广到 n 维立方体和 n 维单纯形.

有趣的是，2009 年在《美国数学月刊》的 533~538 页上，有两位叫 U. Abel 和 M. Ivan 的作者写了一篇文章，题目为《Over-iterates of Bernstein's Operators: A Short and Elementary Proof》，这在 20 年前我们都做过了.

下面的文章原载于:《科学通报》，1986 年，31 卷第 157~160 页.

Limit of Iterates for Bernstein Polynomials Defined on Higher Dimensional Domains

Geng-zhe Chang and Yu-yu Feng

University of Science and Technology of China, Hefei

Received September 26, 1983

1 Introduction

Let $f(x)$ be defined on $[0,1]$. The nth Bernstein polynomials for $f(x)$ are given by

$$B_n(f;x) := \sum_{i=0}^{n} f\left(\frac{i}{n}\right) J_i^n(x), \tag{1}$$

where

$$J_i^n(x) := \binom{n}{i} x^i (1-x)^{n-i}, \quad i = 0, 1, 2, \cdots, n,$$

are called the nth Bernstein basis polynomials. $B_n(f;x)$ is replaced by $B_n(f(x))$ for simplicity.

The iterates for the operator B_n are defined by

$$B_n^{(k)}(f;x) := B_n(B_n^{(k-1)}(f;x)), \quad k = 2, 3, \cdots.$$

In 1967, Kelisky and Rivlin showed in [1] the following theorem: For the arbitrarily fixed n there exists

$$\lim_{k \to \infty} B_n^{(k)}(f;x) = f(0) + [f(1) - f(0)]x. \tag{2}$$

In recent years, different proofs and extensions of (2) have been given by many authors[2-5].

In the present paper, an accurate estimate for the rate of convergence of (2) is presented. The method used to is so simple that we can easily establish some similar theorems for iterates of Bernstein polynomials defined on higher dimensional domains.

2 Limit for iterates of univariate Bernstein polynomials

It is well known that the Berstein operator B_n defined by (1) is linear and positive[6]. Linear functions are invariant under B_n, i.e.

$$B_n(1;x) = 1, \qquad B_n(x;x) = x. \tag{3}$$

Furthermore, we have

$$B_n(X(1-x);x) = \left(1 - \frac{1}{n}\right)x(1-x). \tag{4}$$

Let $[x_1, x_2, \cdots, x_m]f(\cdot)$ denote the divided difference of order $(m-1)$ of function f with respect to $(\cdot)$ at points $x_1, x_2, \cdots, x_m$. We have

Theorem 1 *For the fixed n we set*

$$E_k(f;x) := f(0) + [f(1) - f(0)]x - B_n^{(k)}(f;x).$$

Then

$$|E_k(f;x)| \leqslant \max_{1 \leqslant i \leqslant n-1} \left|\left[0, \frac{i}{n}, 1\right]f(\cdot)\right|\left(1 - \frac{1}{n}\right)^k x(1-x) \downarrow 0 \tag{5}$$

as $k \to \infty$ and equality iff $\left[0, \dfrac{i}{n}, 1\right] f(\cdot)$ is independent of i for $i = 1, 2, \cdots, n-1$.

Proof Inserting $n = 1$ into (1), we get

$$B_1(f;x) = f(0) + [f(1) - f(0)]x. \tag{6}$$

$B_1(f;x)$ can be also expressed by

$$B_1(f;x) = \sum_{i=0}^{n} \left[\left(1 - \frac{i}{n}\right)f(0) + \frac{i}{n}f(1)\right] J_i^n(x),$$

because of (3). Hence

$$\begin{aligned}
E_1(f;x) &= f(0) + [f(1) - f(0)]x - B_n(f;x) \\
&= B_1(f;x) - B_n(f;x) \\
&= \sum_{i=0}^{n} \left[\left(1 - \frac{i}{n}\right)f(0) + \frac{i}{n}f(1) - f\left(\frac{i}{n}\right)\right] J_i^n(x) \\
&= \sum_{i=1}^{n-1} \frac{i(n-i)}{n^2} \left[0, \frac{i}{n}, 1\right] f(\cdot) J_i^n(x).
\end{aligned} \tag{7}$$

It follows that the definition of $E_k(f;x)$,

$$E_k(f;x) = B_n^{(k-1)} \sum_{i=1}^{n-1} \frac{i(n-i)}{n^2} \left[0, \frac{i}{n}, 1\right] f(\cdot) J_i^n(x)$$

$$= B_n^{(k-1)} \sum_{i=0}^{n-2} \left[0, \frac{i+1}{n}, 1\right] f(\cdot) J_i^{n-2} \left(1 - \frac{1}{n}\right) x(1-x).$$

By the linearity and positivity of the operator B_n, and by (4) we obtain

$$|E_k(f;x)| \leqslant \max_{1 \leqslant i \leqslant n-1} \left|\left[0, \frac{i}{n}, 1\right] f(\cdot)\right| \left(1 - \frac{1}{n}\right)^k \times (1-x) \downarrow 0, \qquad (8)$$

as $k \to \infty$. It is obvious that the equality in (8) occurs iff $\left[0, \frac{i}{n}, 1\right] f(\cdot)$ is independent of i for $i = 1, 2, \cdots, n-1$. □

3   Limit of iterates for Bernstein polynomials over triangles

Let T be the triangle with vertices $(0,0), (0,1), (1,0)$. For any function $f(x,y)$, it associates the following polynomials

$$B_n(f;x,y) = \sum_{0 \leqslant i+j \leqslant n} f\left(\frac{i}{n}, \frac{j}{n}\right) J_{i,j}^n(x,y), \qquad (9)$$

which are called the nth Bernstein polynomials for f, where

$$J_{i,j}^n(x,y) = \frac{n!}{i!j!(n-i-j)!} x^i y^j (1-x-y)^{n-i-j}.$$

It is well known that (see [6]) the operator defined by (9) is linear and positive, keeping linear functions in x and y invariant, i.e.

$$B_n(1;x,y) = 1, \quad B_n(x;x,y) = x, \quad B_n(y;x,y) = y, \qquad (10)$$

and having the following properties

$$B_n(x(1-x);x,y) = \left(1 - \frac{1}{n}\right) x(1-x),$$

$$B_n(y(1-y);x,y) = \left(1 - \frac{1}{n}\right) y(1-y), \qquad (11)$$

$$B_n(xy;x,y) = \left(1 - \frac{1}{n}\right) xy.$$

We are now in a position to prove

Theorem 2 *For the fixed n we set*
$$E_k(f;x,y) := xf(1,0) + yf(0,1) + (1-x-y)f(0,0) - B_n^k(f;x,y).$$

Then
$$|E_k(f;x,y)| \leq \left(1-\frac{1}{n}\right)^k [\alpha_n x(1-x) + \beta_n y(1-y) + \gamma_n xy] \downarrow 0 \quad (12)$$

as $k \to \infty$ where
$$\alpha_n := \max_{1\leq i\leq n-1} \left|\left[0,\frac{i}{n},1\right]f(\cdot,0)\right|,$$
$$\beta_n := \max_{1\leq j\leq n-1} \left|\left[0,\frac{j}{n},1\right]f(0,\cdot)\right|,$$
$$\gamma_n := \max_{\substack{i,j\geq 1 \\ i+j\leq n}} \left|\left[0,\frac{i}{n}\right]_x \left[0,\frac{j}{n}\right]_y f(x,y)\right|,$$

in which $\left[0,\dfrac{i}{n}\right]_x$ *means that the divided difference of the first order is taken with respect to x, and* $\left[0,\dfrac{j}{n}\right]_y$ *has the similar meaning. The equality in (12) is attainable.*

Proof We know from the definition (9) that
$$B_1(f;x,y) = f(1,0)x + f(0,1)y + f(0,0)(1-x-y). \quad (13)$$

Straightforward calculation shows that
$$E_k(f;x,y) := B_1(f;x,y) - B_n^{(k)}(f;x,y)$$
$$= B_n^{(k-1)}[B_1(f;x,y) - B_n(f;x,y)]$$
$$= B_n^{(k-1)}\Bigg\{\sum_{i=1}^{n-1} \frac{i(n-i)}{n^2}\left[0,\frac{i}{n},1\right]f(\cdot,0)J_i^n(x)$$
$$+ \sum_{j=1}^{n-1} \frac{j(n-j)}{n^2}\left[0,\frac{j}{n},1\right]f(0,\cdot)J_j^n(y)$$
$$- \sum_{0\leq i+j\leq n} \frac{ij}{n^2}\left[0,\frac{i}{n}\right]_x\left[0,\frac{j}{n}\right]_y f(x,y)J_{i,j}^n(x,y)\Bigg\}.$$

After making further arrangements, we get

$$E_k(f;x,y) = \left(1-\frac{1}{n}\right) B_n^{(k-1)} \Bigg\{ \sum_{i=0}^{n-2} \left[0,\frac{i+1}{n},1\right] f(\cdot,0) J_i^{n-2}(x) \cdot x(1-x)$$

$$+ \sum_{j=0}^{n-2} \left[0,\frac{j+1}{n},1\right] f(0,\cdot) J_j^{n-2}(y) y(1-y)$$

$$- \sum_{0 \leq i+j \leq n-2} \left[0,\frac{i+1}{n}\right]_x \left[0,\frac{j+1}{n}\right]_y f(x,y) J_{i,j}^{n-2}(x,y) xy \Bigg\}.$$

Note that B_n is a linear and positive operator. Using (11) we obtain

$$|E_k(f;x,y)| \leq \left(1-\frac{1}{n}\right)^k [\alpha_n x(1-x) + \beta_n y(1-y) + \gamma_n xy] \downarrow 0, \qquad (14)$$

as $k \to \infty$. The equality occurs when $\left[0,\frac{i}{n},1\right] f(\cdot,0)$ and $\left[0,\frac{j}{n},1\right] f(0,\cdot)$ are positive(negative) constants c_1, c_2 respectively for $i,j = 1,2,\cdots,n-1$, and $\left[0,\frac{i}{n}\right]_x \left[0,\frac{j}{n}\right]_y f(x,y)$ is a negative(positive) constant for positive integers i, j, such that $i+j \leq n$. The proof is completed. □

The last theorem says that the limit of the iterates $B_n^{(k)}(f;x,y)$ is the linear function which interpolates to $f(x,y)$ at the vertices of the triangle T. The theorem can be easily extended to Bernstein polynomials defined on higher dimensional simplices. An analogous theorem for Bernstein polynomials defined on higher dimensional rectangular domains has just been established in [7].

References

[1]   Kelisky R P, Rivlin T J. Iterates of Bernstein Polynomials[J]. Pacific J. of Math, 1967, 21 (3): 511-520.

[2]   Nilson C M, Riesenfeld R F, Weiss W A. Iterates of Markov Operators[J]. J. of Appro Th., 1976, 17 (4): 321-331.

[3]   Karlin S, Ziegler Z. Iteration of Positive Approximation Operators[J]. J. of Appro Th., 1970, 3: 240-249.

[4] 胡莹生，徐叔贤. 一类变差缩减算子的迭代极限[J]. 应用数学学报，1978,1(3)：240-249.

[5] 常庚哲，单墫. A Simple Proof for a Theorem of Kelisky and Rivlin[J]. 数学研究与评论，1983,3 (1): 145-146.

[6] Lorentz G G. Bernstein Polynomials [M]. Toronto: Toronto Univ. Press, 1953.

[7] 冯玉瑜，常庚哲. 定义在矩形上的 Bernstein 多项式的迭代极限[J]. 工程数学报，1985, 1(2):137-141.

11

灯火阑珊觅丽人[①]

电影《居里夫人》中，有这样一个情节：经历了 4 年多的艰辛，在一个大年除夕，居里夫妇满怀希望地等待着镭的出现；可是，当皮埃尔·居里走上前去揭开锅盖时，却什么也没有看到. 4 年的心血几乎是白费了，皮埃尔都想过打退堂鼓. 居里夫人却说："我要知道了失败的原因就不在乎失败了."她在思索，是不是镭所占的比例太少，以至于分离出来的镭不能被肉眼看到？可以说，这个设想实在并不深奥. 我们不说以前的 4 年所取得的成果，耗掉了居里夫妇多大的智慧，只说在最后这个节骨眼上，居里夫人的可贵之处，在于她的执着和追求，在于她敢做"再坚持一下的努力". 最后，在黑暗中她看到了镭在发光，一个震惊世界的奇迹终于出现了.

一个普通的科学工作者，他的成就同居里夫妇的贡献，不可相提并论，但也许都要经历已见胜利的曙光但还有一个险关狭隘的阶段，如果退缩下来，那就前功尽弃，"为山九仞，功亏一篑". 这时只有像居里夫人那样，做再坚持一下的努力，经过几天、几周甚至几个月的思考，或许就在转瞬之间，豁然开朗，回过头来，或许会责备自己"为什么早没有想到呢".

这种体会，我已经有过许多次. 这里，我想把印象最深的那一次介绍给我的青年朋友. 1992 年 11 月下旬，我到瑞典的林雪平大学进行合作研究，时间

[①] 节选自《中国科大报》1998 年 9 月 15 日第 4 版.

只有 4 周. 要想做出一篇能上 SCI 的文章, 时间是太少了. 我在国内时, 早就想好了一个题目, 有过一些准备. 我的合作伙伴, 一个叫拉尔斯, 另一个叫汤姆. 每隔两三天, 我们三人要在一起讨论. 障碍一个个被扫清, 最后归结到一个问题上. 简单地说, 就是怎样确定某个多元函数取负值的点集. 我们花了好长时间, 没有积极的结果. 离开我动身回国的时间只有 7 天了, 仍没有进展. 那天, 拉尔斯在他家为我举行一个 party, 系主任和所有资深教师都到场了, 主人善待于我, 反而使我心情沉重, 我实在不想在合作一无所获的情况下一走了之. 我不得不打起精神, 应酬于瑞典同行之间.

又过了 3 天, 是北欧传统的 Lucia Day (露西亚节), 再有 4 天就要离开林雪平.

瑞典接近北极圈, 但感觉上并不很寒冷, 气温同北京差不多. 在冬季, 白昼特别短, 上午九十点钟天亮, 下午两三点钟就张开了夜幕. 从大学到我的住地, 步行要 40 分钟, 先要穿过一片大森林, 接着要穿过一大片墓地. 瑞典的墓地非常讲究, 一到天黑每座墓前都点起了灯. 可是, 惨淡的灯光反而增添了公墓的阴森恐怖, 每当我从这里走过, 心都紧缩了起来. 我讨厌瑞典的长夜, 看来瑞典人同我一样, 也讨厌长夜, 据说, Lucia Day 是一年中黑夜最长的一天, 过了这一天, 光明就在前面, 所有北欧人都庆祝这个日子, 庆祝活动常在午夜进行.

汤姆换一种方式向我表示告别之意, 他和他的夫人邀请我到另外一个小镇上参加那里的 Lucia 活动. 汽车开了两个小时, 夜里 11 时才到达目的地, 演出还没有开始, 当时的气温已降至零下十几摄氏度, 十分寒冷. 第一个节目给了我最深的印象: 一群少女穿着白色的长裙的小合唱. 站在中央的那位姑娘戴着皇冠, 皇冠上点着好几支蜡烛. 汤姆对我说, 她就是 Lucia, 是光明之神. 她们唱的都是歌颂光明的曲子. 在好客的汤姆夫妇面前, 我只好装出对节目饶有兴致的样子, 其实, 我心烦意乱, 完全没有欣赏的兴致.

凌晨一点时我们启程返回林雪平, 为了让我充分利用时间小睡一下, 汤姆让我一人坐在汽车的后座, 我怎能睡得着! 头脑里还在萦绕着那个问题. 说也奇怪, 就在一瞬之间, 我突然有了新的思路, 随着汽车的高速前进, 虽然没可能做任何计算, 那思路却变得越来越清晰, 而且越来越觉得它是正确的.

当汤姆把我送到我的住所时, 已是凌晨三点多了, 我全然没有睡意, 用了三个小时把整个过程推算了一遍, 确信准确无误, 我眼睁睁地等着天亮. 一到八点钟, 马上拨通拉尔斯和汤姆的电话, 约他们九点钟到办公室同我一起讨论. 我在黑板上详细论证所有的细节，眉飞色舞，兴高采烈，瑞典朋友也认为结果准确无误，皆大欢喜. 后来，形成文字由两位朋友执笔，一直到看校样，都由他们代劳，我不管了. 文章很长，印成杂志有 21 页.

1992 年 12 月 22 日，我登上了返回中国的飞机，心情特别愉快. 从斯德哥尔摩起飞，在芬兰的赫尔辛基换机，经曼谷来到香港，同妹妹一家欢度圣诞平安夜.

王国维用词的形式刻画了治学的三种境界. 第三种境界是："众里寻她千百度，蓦然回首，那人却在灯火阑珊处." 这些话似乎有国际意义，不管是对中国人，还是对波兰人、法国人……都是能适用的.

下面的文章原载于：《Linear Algebra and its Applications》，1995 年，220 卷第 9~30 页.

Criteria for Copositive Matrices Using Simplices and Barycentric Coordinates

Lars-Erik Andersson[①]   Geng-zhe Chang[②]   and   Tommy Elfving[③]

Abstract: We present criteria for verifying the copositivity of an $n \times n$ matrix, given that all its principal submatrices of order $n-1$ are copositive. For $n = 4, 5$ necessary and sufficient conditions for copositivity are given, based on the sign distribution of the off-diagonal elements of a single row. Here it is only assumed that one principal submatrix is copositive.

1 Introduction

Let S_n be the set of real and symmetric matrices of order n. The matrix $A \in S_n$ is called copositive (cop) if it belongs to the closed convex cone C, with

$$C = \{A \in S_n : x^\mathrm{T} A x \geqslant 0, \quad \forall x \in \mathbf{R}_+^n\}. \tag{1}$$

Obviously the cone of positive semidefinite matrices, C_1, and the cone of elementwise nonnegative matrices, C_2, both belong to C. For $n = 2$ one has $C = C_1 \cup C_2$ and for $n = 3, 4$ it holds that $C = C_1 + C_2$. However, for $n > 4$, $C \neq C_1 + C_2$, cf. [7]

Since Motzkin[13] introduced the concept of copositive matrices, many new results and generalizations have appeared. The characterization of copositive matrices is treated e.g. in [4], [5], [6] (which also contains a short but very instructive survey), [11], [12] and [16]. Several authors also treat more general

---

① Department of Mathematics, Linköping University, S-581 83 Linköping, Sweden.

② Department of Mathematics, University of Science and Technology of China, Hefei, Anhui 230026, China.

③ Department of Mathematics, Linköping University, S-581 83 Linköping, Sweden.

cases. e.g. when the vector x belongs to a convex polytope or to some other closed convex set; see [10], [11], [12] and [17]. The problem of characterizing the extremal rays of C is described e.g. in the book by Hall[7]. Some other references with a combinatorial approach are [1], [2], and [8]. In [14] it is shown that testing whether a given integer square matrix is not copositive is NP-complete.

There is a simple connection between the copositivity of matrices and the nonnegativity of simplicial Bernstein-Bézier quadratic functions. We recall the notion of a simplex of dimension $n-1$. Let $V = \{V_1, V_2, \cdots, V_n\}$ be n given points in some vector space $\boldsymbol{V}$ such that the $n-1$ vectors $\overrightarrow{V_1V_2}, \overrightarrow{V_1V_3}, \cdots, \overrightarrow{V_1V_n}$ are linearly independent. Then $[V]$ is called an $n-1$-simplex (or a simplex of dimension $n-1$) with respect to $\{V_i\}$, where

$$[V] = \left\{ \boldsymbol{z} \in \boldsymbol{V} : \boldsymbol{z} = \sum_1^n u_i V_i, \ \sum_1^n u_i = 1, \ u_i \geqslant 0 \right\}.$$

We may alternatively represent $[V]$ as

$$U = \left\{ \boldsymbol{u} \in \mathbf{R}^n : \boldsymbol{u} = (u_1, u_2, \cdots, u_n)^T, \ \sum_1^n u_i = 1, \ u_i \geqslant 0 \right\}.$$

The components, $u_i = u_i(z)$, of the vector $\boldsymbol{u}$ are called the barycentric coordinates of the point $\boldsymbol{z}$ with respect to $[V]$. It is clear that V_1 is represented by $(1, 0, \cdots, 0)$, V_2 by $(0, 1, \cdots, 0)$, etc. A simplex of dimension 1 is a segment in $\boldsymbol{V}$. A simplex of dimension 2 is a triangle in $\boldsymbol{V}$, a 3-dimensional simplex is a tetrahedron, and so forth.

The quadratic form

$$p(\boldsymbol{u}) := \boldsymbol{u}^{\mathrm{T}} A \boldsymbol{u}, \quad \boldsymbol{u} \in U, \tag{2}$$

is said to be a quadratic Bernstein-Bézier (surface) patch over the $n-1$-dimensional simplex $[V]$. The Bernstein-Bézier patches of degree k over triangles have been widely investigated in computer-aided geometric design (CAGD). In shape-preserving approximation one often needs to impose restrictions on the approximation, e.g. nonnegativity. For more motivation and details on this problem we refer to the work of Nadler[15] and Chang and Sederberg[3].

We observe the simple fact that the nonnegativity of $p(\boldsymbol{u})$ on $[V]$ is equivalent to the copositivity of the coefficient matrix A, i.e.,

$$\boldsymbol{u}^{\mathrm{T}} A \boldsymbol{u} \geqslant 0, \quad \boldsymbol{u} \in U \quad \Leftrightarrow \quad A \in C. \tag{3}$$

This was pointed out by Micchelli and Pinkus in [12], where also an iterative procedure is proposed, based on the Bernstein-Bézier representation, to test if a polynomial (of degree k in $\mathbf{R}^n$) is positive on a simplex. In this paper we will use the relation (3) when analyzing copositivity.

In Section 2 we establish criteria for the copositivity of an $n \times n$ matrix, given that all its principal submatrices of order $n-1$ are copositive. We also summarize our results in a recursive algorithm for testing whether a given matrix is copositive or not. This algorithm might be useful for small values of n.

Quite recently Li and Feng[9] determined all copositive matrices in S_4. They considered the sign distribution of the six off-diagonal elements. In Section 3 we will present a somewhat simpler analysis, for matrices in S_4 and S_5, where we consider the sign distribution of the off-diagonal elements in a single row. The method in Section 3 is quite different from and more explicit than the one in Section 2. The main difference is that in Section 3 we only assume that one principal submatrix is known to be copositive. The results in Section 3 are derived using an equivalence between copositivity and the subdivision of a particular solid in $\mathbf{R}^{n-1}$ into nonintersecting simplices.

2 Criteria for $n \times n$ matrices

A quadratic Bernstein-Bézier polynomial on the interval $[0, 1]$ is defined by

$$p(x) \doteq a_{11}(1-x)^2 + 2a_{12}(1-x)x + a_{22}x^2,$$

with $x \in [0, 1]$. We have the following

Lemma 1 $p(x) \geqslant 0$ for all $x \in [0,1]$ if and only if the inequalities

$$a_{11} \geqslant 0, \quad a_{22} \geqslant 0, \quad \sqrt{a_{11}a_{22}} + a_{12} \geqslant 0 \tag{4}$$

hold simultaneously.

Proof Suppose first that $p(x) \geqslant 0$ for all $x \in [0,1]$. It follows that $a_{11} = p(0) \geqslant 0$ and $a_{22} = p(1) \geqslant 0$. Further we may write

$$p(x) = [(1-x)\sqrt{a_{11}} - x\sqrt{a_{22}}]^2 + 2x(1-x)[\sqrt{a_{11}a_{22}} + a_{12}]. \tag{5}$$

If one of a_{11} and a_{22} is zero, then (5) implies at once $a_{12} \geqslant 0$. Suppose now that $a_{11} > 0$ and $a_{22} > 0$. Let

$$x^* \doteq \frac{\sqrt{a_{11}}}{a_{11} + \sqrt{a_{22}}}.$$

It is clear that $x^* \in (0,1)$. From (5) we see that

$$p(x^*) = 2x^*(1-x^*)[\sqrt{a_{11}a_{22}} + a_{12}] \geqslant 0 \quad \Rightarrow \quad \sqrt{a_{11}a_{22}} + a_{12} \geqslant 0.$$

This completes the proof of necessity.

Next suppose that the conditions (4) are satisfied. It follows immediately from (5) that $p(x) \geqslant 0$ for all $x \in [0,1]$. □

This simple lemma, also given by Nadler[15] (but with a more involved proof), will be repeatedly quoted in the sequel.

We now recall the following useful result:

Lemma 2 *If A is copositive, so is any principal submatrix of A, any symmetric permutation of A, and any matrix of the form DAD where D is a diagonal matrix with nonnegative diagonal elements.*

We shall also frequently use the following simple but for our purposes important result:

Lemma 3 *Let $a \geqslant 0$ and $b \leqslant 0$. Then*

$$\sqrt{a} + b \geqslant 0 \quad \Leftrightarrow \quad a - b^2 \geqslant 0.$$

Next we present

Theorem 4 *Let $A \in S_n$ be partitioned as*

$$A = \begin{pmatrix} a_{11} & \bar{a}_1^T \\ \bar{a}_1 & A_2 \end{pmatrix},$$

and define the matrix $B \in S_{n-1}$ as

$$B = a_{11}A_2 - \bar{a}_1\bar{a}_1^T, \quad \bar{a}^T = (a_{12}, a_{13}, \cdots, a_{1n}).$$

Assume that

$$a_{11} \geqslant 0, \quad A_2 \text{ cop.} \tag{6}$$

Then the following holds:

(1) If $\bar{a}_1 \geqslant 0$, then A is cop.

(2) If $\bar{a}_1 \leqslant 0$, then A is cop $\Leftrightarrow$ B is cop.

Proof We introduce the standard simplex in $\mathbf{R}^{n-1}$,

$$T = \left\{ \bar{u} \in \mathbf{R}^{n-1} : \bar{u} = (u_2, u_3, \cdots, u_n)^T \geqslant 0, \sum_{2}^{n} u_i = 1 \right\}.$$

The sets U and T are illustrated for $n=4$ in Figure 1. From (3), with $u = (1-t, t\bar{u}) \in U$,

$$p(1-t, t\bar{u}) \geqslant 0, \quad \bar{u} \in T, \quad 0 \leqslant t \leqslant 1 \quad \Leftrightarrow \quad A \text{ is cop.}$$

[Here we write, for ease of notation, $u = (1-t, t\bar{u})$ instead of $u = (1-t, t\bar{u}^T)^T$.] By expansion,

$$p(1-t, t\bar{u}) = (1-t)^2 a_{11} + 2t(1-t)\bar{a}_1^T \bar{u} + t^2 p(0, \bar{u}).$$

Now $p(0, \bar{u}) = \bar{u}^T A_2 \bar{u}$, and by the assumption (6) and Lemma 1,

$$A \text{ is cop} \quad \Leftrightarrow \quad \varphi \doteq \sqrt{a_{11} p(0, \bar{u})} + \bar{a}_1^T \bar{u} \geqslant 0, \quad \bar{u} \in T.$$

Hence case 1 follows.

Next we introduce the sets

$$T^+ = \{\bar{u} : \bar{u} \in T \text{ and } \bar{a}_1^T \bar{u} \geqslant 0\}, \quad T^- = T \setminus T^+.$$

Let
$$\psi = a_{11}p(0,\overline{u}) - (\overline{a}_1^T \overline{u})^2 \equiv \overline{u}^T B \overline{u}. \tag{7}$$

By Lemma 3,
$$\forall \,\overline{u} \in T^-, \quad \varphi(\overline{u}) \geqslant 0 \quad \Leftrightarrow \quad \psi(\overline{u}) \geqslant 0, \tag{8}$$

i.e.,
$$A \text{ is cop} \quad \Leftrightarrow \quad \overline{u}^T B \overline{u} \equiv \psi(\overline{u}) \geqslant 0 \quad \forall \,\overline{u} \in T^-. \tag{9}$$

However, for the case 2 the set T^+ has empty (relative) interior and thus the closure of T^- is T. □

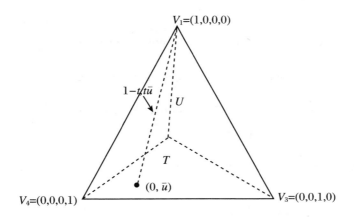

Fig 1 The sets U and T for $n = 4$

Note that if $a_{11} \neq 0$ then $(1/a_{11})B$ equals the Schur complement of the matrix A w.r.t. the partitioning in Theorem 4.

We introduce the notation
$$\text{psm}(A, k)$$

for any principal submatrix of A of order k. We will now derive results for the case when not all elements of $\overline{a}_1$ have the same sign. In order to obtain these results we well replace (6) with stronger conditions:

$$\text{All psm}(A, n-1) \text{ are cop.} \tag{10}$$

Then we first have the following result.

Lemma 5 *Assume (10). If A is noncopositive (ncp), then all $psm(A, n-1)$ are positive semidefinite.*

Proof If A is ncp, then the quadratic form $p(\boldsymbol{u})$ has a negative minimum at some interior point of U. By (10) $p \geqslant 0$ on each face $u_i = 0$. Further, the restriction of p to a straight line is either convex or concave. We conclude that the restriction of p to any straight line through the interior minimizing point is convex. It follows that $p(\boldsymbol{u}) \geqslant 0$ for $\boldsymbol{u} \notin U$. In particular, $p \geqslant 0$ if $u_i = 0$ and $\sum_{j \neq i} u_j = 1$, with no sign restriction on $u_j, j \neq i$. Hence also $p \geqslant 0, u_i = 0, \forall u_1, u_2, \cdots, u_{i-1}, u_{i+1}, \cdots, u_n$. □

We next note, by (9), the following equivalence:

$$A \text{ is ncp} \quad \Leftrightarrow \quad \psi(\overline{u}) < 0 \text{ for some } \overline{u} \in T^-. \tag{11}$$

In the following lemma ∂T^- denotes the topological boundary of T^- considered as a subset of $\mathbf{R}^{n-2}$ and not of $\mathbf{R}^{n-1}$.

Lemma 6 *Assume (10). Then $\psi(\overline{u}) \geqslant 0$ on ∂T^-.*

Proof By assumption, $p(1-t, t\overline{u}) \geqslant 0, t \in [0,1]$, whenever some $u_i = 0, i > 1$. Hence by Lemma 1, $\varphi(\overline{u}) \geqslant 0, \overline{u} \in T$, whenever some $u_i = 0$. If in addition $\overline{u} \in T^-$, then by (8), $\psi(\overline{u}) \geqslant 0$. Hence $\psi \geqslant 0$ on $\partial T^- \setminus L_0$ with

$$L_0 = \left\{ \overline{u} \in \mathbf{R}^{n-1} : \overline{u} \geqslant 0, \ \sum_2^n u_i = 1, \ \overline{a}_1^{\mathrm{T}} \overline{u} = 0 \right\}.$$

On L_0 we have $\psi = a_{11} p(0, \overline{u}) \geqslant 0$ by (7). □

Lemma 7 *Assume that all $psm(A, n-1)$ are positive semidefinite. Then all $psm(B, n-2)$ are positive semidefinite.*

Proof Let $\overline{u}_i = (u_2, u_3, \cdots, u_{i-1}, 0, u_{i+1}, \cdots, u_n)^{\mathrm{T}}$. Then

$$p(1-t, t\overline{u}_i) \equiv (1-t, t\overline{u}_i)A(1-t, t\overline{u}_i)^{\mathrm{T}} \geqslant 0 \quad \text{for all } t, \overline{u}_i.$$

Hence also $p(1-t, -t\overline{u}_i) \geqslant 0$ for all $t, \overline{u}_i$. It follows form Lemma 1 that

$$\sqrt{a_{11}p(0, \overline{u}_i)} \pm \overline{a}_1^{\mathrm{T}} \overline{u}_i \geqslant 0 \quad \text{for all } \overline{u}_i,$$

i.e., that $\psi(\overline{u}_i) = \overline{u}_i^{\mathrm{T}} B \overline{u}_i \geqslant 0$. □

By Lemma 5 we immediately get the following corollary.

Corollary 8 *Assume (10). If A is ncp, then all $psm(B, n-2)$ are positive semidefinite.*

Now let us assume that the coordinates $x_2, x_3, \cdots, x_n$, have been reordered so that for the vector $\overline{a}_1$ in the matrix A, we have

$$a_{12} \geqslant a_{13} \geqslant \cdots \geqslant a_{1n}.$$

Then apart from cases 1 and 2 in Theorem 4 we have only the following two possible sign distributions:

case 3a: $\quad a_{12} > 0, \quad a_{1n} < 0, \quad a_{12} > a_{13} \geqslant a_{14} \geqslant \cdots \geqslant a_{1n},$

and

case 3b: $\quad a_{12} > 0, \quad a_{1n} < 0,$
$$a_{12} = a_{13} = \cdots = a_{1i} > a_{1i+1} \geqslant \cdots \geqslant a_{1n}, \quad i > 2.$$

This ordering of the elements is assumed for ease of presentation only and is not necessary. Without this assumption the matrices D and W below will be replaced by symmetric permutations.

To treat case 3 we shall need some further concepts. Let

$$S^+ = \left\{ \overline{u} \in \mathbf{R}^{n-1} : u_3, u_4, \cdots, u_n \geqslant 0, \sum_2^n u_i = 1, \overline{a}_1^{\mathrm{T}} \overline{u} \geqslant 0 \right\}.$$

For case 3a consider the points $\{\overline{V}_i\}_2^n \in S^+$,

$$\overline{V}_2 = (1, 0, \cdots, 0)^{\mathrm{T}}, \quad \overline{V}_i = (u_2^i, 0, \cdots, u_i, 0, \cdots, 0)^{\mathrm{T}}, \quad i > 2,$$

where u_i is in the $(i-1)$th position in $\overline{V}_i$ and such that for $i > 2$ it holds that $\overline{a}_1^T \overline{V}_i = 0$ and $u_2^i + u_i = 1$. This implies

$$u_2^i = \frac{-a_{1i}}{a_{12} - a_{1i}}, \quad u_i = \frac{a_{12}}{a_{12} - a_{1i}}.$$

It is easy to verify that the vectors $\{\overrightarrow{V_2 V_i}\}_{i>2}$ are linearly independent and hence $\{\overline{V}_i\}_2^n$ generates a simplex in $\mathbf{R}^{n-1}$ which is identical to the set S^+.

Let $\alpha = (\alpha_2, \cdots, \alpha_n)$ be the barycentric coordinates w.r.t. S^+. Then (remember that $\overline{u} \in \mathbf{R}^{n-1}$ are the barycentric coordinates w.r.t. T, the standard simplex in $\mathbf{R}^{n-1}$)

$$\overline{u}^T = \alpha^T W,$$

where the ith row of the $(n-1) \times (n-1)$ matrix W equals $\overline{V}_{i+1}$. Note that W is left triangular and nonsingular.

Define

$$S^- = \left\{ \overline{u} = W^T \alpha \in \mathbf{R}^{n-1} : \sum_2^n \alpha_j = 1, \; \alpha_2 < 0, \; \alpha_3, \alpha_4, \cdots, \alpha_n \geqslant 0 \right\}.$$

Then

$$S^- \cap T = T^- \quad \text{and} \quad S^+ \cap T = T^+.$$

Now

$$\psi(\overline{u}) \equiv \overline{u}^T B \overline{u} = \alpha^T W B W^T \alpha.$$

Then [using (11)].

Lemma 9 *Assume (10) and case 3a. Then*

$$\psi(\overline{u}) < 0 \; for \; some \; \overline{u} \in S^- \quad \Leftrightarrow \quad DWBW^T D \; is \; ncp,$$

with D a diagonal matrix, $d_1 = -1, d_i = 1, i > 1$.

For case 3b we may obtain a similar result, e.g. by taking limits in case 3a: First note that the copositivity of the matrix $DWBW^T D$ is not affected if we multiply W by a constant. If $a_{12} > a_{13} \geqslant a_{14} \geqslant \cdots \geqslant a_{1i} > a_{1i+1} \geqslant \cdots \geqslant a_{1n}$, then multiply the previous matrix W by the positive number $a_{12} - a_{1i}$. Then let $a_{1j} \to a_{12}, j = 3, 4, \cdots, i$, keeping $a_{1i+1}, \cdots, a_{1n}$ fixed. The resulting matrix $\overline{W}$ has all rows $\overline{V}_j$ equals to zero except for $j = 2, 3, \cdots, i$. We have $\overline{V}_j = (-a_{12}, 0, \cdots, a_{12}, 0, \cdots)$ with the nonzero elements in the 1st and $(j-1)$st positions $(j = 2, 3, \cdots, i)$. The statement in Lemma 9 is now valid with W replaced by $\overline{W}$.

The set S^+, S^-, etc. are illustrated in Figures 2 (case 3a) and 3 (case 3b) for $n = 4$. For case 3a, S^+ equals the triangle with corners at $\overline{V}_2, \overline{V}_3, \overline{V}_4$, whereas S^- equals the unbounded quadrilateral with two of its corners at $\overline{V}_3$ and $\overline{V}_4$ and lying between the rays $u_3 = 0$ and $u_4 = 0$.

We are now ready for

Theorem 10 *Assume (10). Then the following results holds:*

$$A \text{ is ncp} \Leftrightarrow B \text{ and } DWBW^TD(D\overline{W}B\overline{W}^TD) \text{ are ncp}$$

$$\text{and all } psm(A, n-1) \text{ are positive semidefinite.}$$

Alternatively this theorem can be rephrased as

Theorem 11 *Assume (10). Then the following holds:*

$$A \text{ is ncp} \Leftrightarrow B \text{ or } DWBW^TD(D\overline{W}B\overline{W}^TD) \text{ is cop}$$

$$\text{or some } psm(A, n-1) \text{ is not positive semidefinite.}$$

Proof (*Proof of Theorem 10*). Suppose first that A is ncp. Then, by (11), $\overline{u}^T B \overline{u} < 0$ for some $\overline{u} \in T^-$, which implies that B is ncp. Further, it is clear from Lemma 9 that (since $T^- \subset S^-$) $DWBW^TD$ is ncp. By Lemma 5 it also follows that all $psm(A, n-1)$ are positive semidefinite.

Conversely, assume that B and $DWBW^TD$ are ncp and that all $psm(A, n-1)$ are positive definite. Then

$$\overline{u}_0^T B \overline{u}_0 < 0 \quad \text{for some } \overline{u}_0 \in T$$

and (Lemma 9)

$$\overline{u}_1^T B \overline{u}_1 < 0 \quad \text{for some } \overline{u}_1 \in S^-.$$

By Lemma 6 it follows that $\overline{u}^T B \overline{u} \geqslant 0$ when $\overline{u} \in \partial T^-$, and by Lemma 7 that all $psm(B, n-2)$ are positive semidefinite. Now connect $\overline{u}_0$ and $\overline{u}_1$ by a straight line L (cf. Fig 4), where also the points on L where the restriction of $\bar{u}^T B \bar{u}$ (to L) is nonnegative are marked with a $+$ sign. Since the restriction is quadratic, it follows that $\bar{u}_0 \in T^- = S^- \cap T$ and $\bar{u}_1 \in T^-$. In particular it follows that $\bar{u}^T B \bar{u} < 0$ for some $\bar{u} \in T^-$, which, by (11), implies that A is ncp. This completes the proof. □

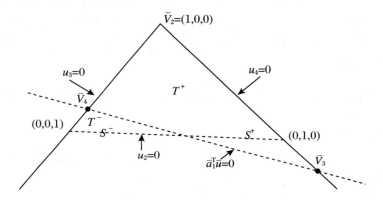

Fig 2 Case 3a, $n = 4$ and $a_{13} > 0$

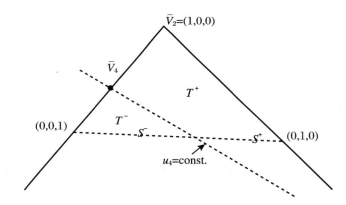

Fig 3 Case 3b, $n = 4$ and $a_{13} > 0$

Remark 2.1 *A similar result was given by Cottle, Habetler, and Lemke in [5, Theorem 3.1]. Here the equivalent conditions are det $A < 0$ and adj $A \geqslant 0$, where adj A is the adjoint matrix (i.e. the matrix of cofactors of A).*

Using the above theorems, we may now formulate the following *recursive*

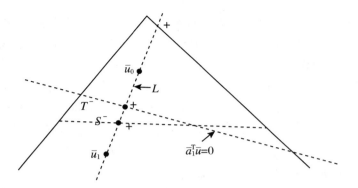

Fig 4   Illustration of Theorem 10 for $n = 4$

algorithm for deciding whether a matrix $A \in S_n$ is copositive. Note again that the coordinates $x_2, \cdots, x_n$ are assumed to be reordered so that we have the previously mentioned inequalities $a_{12} \geqslant a_{13} \geqslant \cdots \geqslant a_{1n}$ for the vector $\bar{a}_1$.

(1) *First let $A(k) = psm(A, k)$ be any of the* $\binom{n}{k}$ *principal submatrices of order k of A. Also let*

$$A(k) = \begin{pmatrix} a_{11} & \bar{a}_1^{\mathrm{T}} \\ \bar{a}_1 & A_2 \end{pmatrix},$$

where the index k is suppressed in the partitioning. Put $k = 2$.

(2) *Next every $A(k)$ is tested for copositivity in the following way.*

2a. *Test if some $A(k-1)$ is not positive semidefinite. If so, then $A(k)$ is cop. Otherwise go to 2b.*

2b. *The condition (6) of Theorem 4 is now valid for $A = A(k)$. Test if $\bar{a}_1 \geqslant 0$. If so, then $A(k)$ is cop (case 1 of Theorem 4). Otherwise go to* 2c.

2c. *If $\bar{a}_1 \leqslant 0$ go to* 2d. *Otherwise go to* 2e.

2d. *Test the matrix $B(k) = a_{11}A_2 - \bar{a}_1\bar{a}_1^{\mathrm{T}}$ (of order $k-1$) for copositivity, and use $B(k)$ is cop $\Leftrightarrow A(k)$ is cop (case 2 of Theorem 4).*

2e. *Test the matrix $B(k)$ for copositivity. If $B(k)$ is cop then $A(k)$ is cop (Theorem 11). If $B(k)$ is ncp, then go to* 2f.

2f. *Test the matrix $DWB(k)W^{\mathrm{T}}D$ (of order $k-1$) for copositivity, and use $DWB(k)W^{\mathrm{T}}D$ is cop $\Leftrightarrow A(k)$ is cop (Theorem 11).*

3. *If all psm(A, k) have been found to be cop, then take $k := k+1$ and go to 2.*

In steps 2d—2f above, the matrices $B(k)$ and $DWB(k)W^T D$ are tested for copositivity. We then note the following important facts.

When testing the matrix $B(k)$, we already know, from the previous step 2a of the algorithm, that all $A(k-1)$ [i.e. all $psm(A(k), k-1)$] are positive semidefinite. By Lemma 7 we then conclude that all $psm(B(k), k-2)$ are positive semidefinite. Therefore, if we reiterate the algorithm with $A(k)$ replaced by $B(k)$, we may go directly to step 2b and thereby avoid the time-consuming tests for copositivity and definiteness of all the $psm(B(k), k-2)$.

A similar observation can be made for the matrix $DWB(k)W^T D$ in step 2f. For ease of notation we suppress the k dependence of the matrices D and W. By Lemma 12 below, this matrix also has the property that all its psm of order $k-2$ are positive semidefinite.

Lemma 12 *If all $psm(A(k), k-1)$ and all $psm(B(k), k-2)$ are positive semidefinite, then all $psm(DWB(k)W^T D, k-2)$ are positive semidefinite.*

Proof Let $\overline{u} \in \mathbf{R}^{k-1}$. Using the previous notation, we have
$$\psi(\overline{u}) = a_{11} p(0, \overline{u}) - (\overline{a}_1^T \overline{u})^2 \equiv \overline{u}^T B(k) \overline{u}.$$
Now $\overline{u}^T = \alpha^T W$ $[\alpha = (\alpha_2, \alpha_3, \cdots, \alpha_k)^T]$. Hence
$$\psi(\overline{u}) = \alpha^T W B(k) W^T \alpha.$$
By the construction of the set S^+,
$$\overline{a}_1^T \overline{u} = 0 \quad \Leftrightarrow \quad \alpha_2 = 0.$$
Since all $psm(A(k), k-1)$ are positive semidefinite, it follows that
$$a_{11} p(0, \overline{u}) = (0, \alpha_3, \cdots, \alpha_k) DWB(k) W^T D (0, \alpha_3, \cdots, \alpha_k)^T \geqslant 0$$
for all $\overline{u}$, i.e. for all $(\alpha_3, \alpha_4, \cdots, \alpha_k)$.

Now for case 3a, the matrix W is left triangular and nonsingular. It follows, for $i > 2$, that the subspace $\{\overline{u} : u_i = 0\}$ is mapped onto the subspace $\{\alpha : \alpha_i = 0\}$ by the transformation $\overline{u} = \alpha W$. Hence, for $i > 2$ and with $u_i = 0$,
$$0 \leqslant \overline{u}^T D B(k) D \overline{u} = \alpha^T DWB(k) WD\alpha,$$

and we conclude that $DWB(k)WD$ has all its psm of order $k-2$ positive semidefinite. For case 3b the previous argument is easily modified (this is, however, omitted). □

3 Criteria for $n = 4$ and $n = 5$

The conditions in (4) are both necessary and sufficient for the copositivity of matrices in S_2. We now review the copositivity of matrices in S_3. In 1983, Hadeler[6] found all copositive matrices in S_3. He showed that $A \in S_3$ is copositive if and only if the inequalities

$$a_{ii} \geqslant 0, \quad i = 1, 2, 3, \tag{12}$$

$$\bar{a}_{ij} \doteq \sqrt{a_{ii}a_{jj}} + a_{ij} \geqslant 0, \quad i \neq j, \; i,j = 1,2,3, \tag{13}$$

are satisfied, as well as at least one of the following conditions:

$$\bar{a} \doteq a_{23}\sqrt{a_{11}} + a_{31}\sqrt{a_{22}} + a_{12}\sqrt{a_{33}} + \sqrt{a_{11}a_{22}a_{33}} \geqslant 0, \tag{14}$$

$$det A \geqslant 0. \tag{15}$$

In 1992, when characterizing the nonnegativity of Berstein-Bézier triangular patches, Nadler[15] obtained the same results by a different approach. Chang and Sederberg[3] recently gave a short proof of these results and pointed out that the last two inequalities can be written equivalently as the following single inequality:

$$\bar{a} + \sqrt{2\bar{a}_{12}\bar{a}_{13}\bar{a}_{23}} \geqslant 0. \tag{16}$$

Quite recently Li and Feng[9] determined all copositive matrices in S_4. Their results are displayed by case analysis. They considered the sign distribution of the six off-diagonal elements. We will present a somewhat simpler analysis where we consider the sign distribution of the three off-diagonal elements in a single row (say row 1). We will then always assume that (6) holds, but not necessarily (10).

We will first show that copositivity is closely related to the subdivision of the set T^-, introduced in Section 2, into nonintersecting simplices. Let

$mat\{\nu_1, \nu_2, \cdots, \nu_i\}$ denote a matrix with rows $\nu_1, \nu_2, \cdots, \nu_i$ taken in any order. The order will be irrelevant when testing for copositivity, due to Lemma 2. Further, $vert(T^-)$ denotes the set of vertices of T^-.

Lemma 13 *Assume that T^- can be subdivided into I simplices S^i in $\mathbf{R}^{n-1}$ such that $T^- = \cup_{i=1}^I S^i, S^i \cap S^j \neq \varnothing$ is a subsimplex of S^i and S^j if $i \neq j$ and $vert(S^i) \subset vert(T^-)$. Let $vert(S^i) = \{V_j^i\}_{j=1}^{n-1}$, and define $W^i = mat\{V_1^i, V_2^i, \cdots, V_{n-1}^i\}$ (here V_j^i is represented in the barycentric coordinates of T).*

Let $A \in S_n$ be partitioned as

$$A = \begin{pmatrix} a_{11} & \bar{a}_1^T \\ \bar{a}_1 & A_2 \end{pmatrix},$$

and define the matrix $B \in S_{n-1}$ as

$$B = a_{11} A_2 - \bar{a}_1 \bar{a}_1^T, \quad \bar{a}_1^T = (a_{12}, a_{13}, \cdots, a_{1n}).$$

Assume that

$$a_{11} \geqslant 0, \quad A_2 \text{ cop.}$$

Then the following holds:

$$A \text{ is cop} \quad \Leftrightarrow \quad W^i B (W^i)^T, i = 1, 2, \cdots, I, \text{ are cop.}$$

Proof For each simplex S^i let $\alpha = (\alpha_1, \alpha_2, \cdots, \alpha_{n-1})^T$ be its barycentric coordinates (for ease of notation we suppress the dependence on i in α). Recall that $\bar{u}^T = (u_2, u_3, \cdots, u_n)$ are the barycentric coordinates with respect to T. Then

$$\bar{u}^T = \alpha^T W^i.$$

Hence

$$\bar{u} \in S^i \quad \Leftrightarrow \quad (W^i)^T \alpha \in S^i, \quad \sum_1^{n-1} \alpha_l = 1, \quad \alpha \geqslant 0.$$

It follows, using (9), that

$$A \text{ is cop} \quad \Leftrightarrow \alpha^T W^i B (W^i)^T \alpha \geqslant 0, \quad \sum_1^{n-1} \alpha_l = 1, \alpha_l \geqslant 0, i = 1, 2, \cdots, I.$$

Note that only the set T^- needs to be inspected (since $\bar{u}^T A \bar{u} \geq 0$ holds trivially for $\bar{u} \in T^+$). □

We now introduce some new notation for the purpose of describing the vertices of the solid T^-. First let the vertices of the simplex T be denoted $\{T_i\}_1^{n-1}$, where each T_i can be represented in the barycentric coordinates of T as (e_i is the ith row of the identity matrix),

$$T_i = e_i \in \mathbf{R}^{n-1}.$$

We first derive the intersection between the edges of T and the hyperplane $\bar{a}_1^T \bar{u} = 0$. Assume that $a_{1i} \neq a_{1j}$. The equation $a_{1i} u_i + a_{1j} u_j = 0, u_i + u_j = 1$ will then have the solution

$$\hat{u}_j = -\frac{-a_{1i}}{a_{1j} - a_{1i}}, \quad \hat{u}_i = \frac{a_{1j}}{a_{1j} - a_{1i}}.$$

The intersection point along the edge from T_i to T_j, $\hat{V}^{i,j} \in \mathbf{R}^{n-1}$, becomes

$$(\hat{V}^{i,j})_1 = \begin{cases} \hat{u}_{i+1}, & l = i, \\ \hat{u}_{j+1}, & l = j, \\ 0, & \text{else}. \end{cases}$$

Note that $\hat{V}^{ij} = \hat{V}^{ji}$. For later use we also define

$$(\overline{V}^{i,j})_l = \begin{cases} a_{1,j+1}, & l = i, \\ -a_{1,i+1}, & l = j, \\ 0, & \text{else}. \end{cases}$$

Here $\overline{V}^{ij} = -\overline{V}^{ji}$.

Let $\{i_1, i_2, \cdots, i_k\}$ be a subset of $\{2, 3, \cdots, n\}$ with distinct elements. Assume that exactly k elements are negative in $\bar{a}_1$, say $a_{1i_1}, a_{1i_2}, \cdots, a_{1,i_k} < 0$. Let $vert(T^-)$ denote the set of vertices of T^-. Note that $a_{1j} < 0 \Leftrightarrow T_{j-1} \in vert(T^-)$. Hence the set of vertices of T^- includes

$$\overline{T}(k) = \{T_{i_l - 1}\}_{l=1}^k.$$

The other vertices of T^- can be described as follows. For each $i_l, l = 1, 2, \cdots, k$, the solid T^- has vertices along all edges going from $T_{i_l - 1}$ to T_i (this is vertex

$\hat{V}^{i_l-1,i}$) with $T_i \notin \overline{T}(k)$. We will define this formally:

$$N_{i_l} = \{\hat{V}^{i_l-1,i} : T_i \notin \overline{T}(k)\}, \quad l = 1, 2, \cdots, k.$$

Then the set of vertices of T^- is

$$vert(T^-) = \{\overline{T}(k), \{N_{i_l}\}_{l=1}^k\}.$$

It is easy to verify that

$$\text{card}(vert(T^-)) = k(n-k).$$

We may now formulate the main result of this section.

Theorem 14 *Let $A \in S_n$ be partitioned as*

$$A = \begin{pmatrix} a_{11} & \overline{a}_1^T \\ \overline{a}_1 & A_2 \end{pmatrix},$$

and define the matrix $B \in S_{n-1}$ as

$$B = a_{11}A_2 - \overline{a}_1\overline{a}_1^T, \quad \overline{a}_1^T = (a_{12}, a_{13}, \cdots, a_{1n}).$$

Assume that

$$a_{11} \geqslant 0, \quad A_2 \text{ cop.} \tag{17}$$

Then the following hold:

(1) If $\overline{a}_1 \geqslant 0$, then A is cop.

(2) If $\overline{a}_1 \leqslant 0$, then A is cop $\Leftrightarrow B$ is cop.

(3) If exactly on element a_{1,i_1+1} is negative, then A is cop $\Leftrightarrow W(i_1)BW(i_1)^T$ is cop. Here

$$W(i_1) = mat\{e_{i_1}, \overline{V}^{i_1,j_1}, \overline{V}^{i_1,j_2}, \cdots, \overline{V}^{i_1,j_{n-2}}\}, \quad j_l \in \{1, 2, \cdots, n-1\} \setminus \{i_1\}.$$

and $j_l \neq j_{l'}, l \neq l'$.

(4) Let $n = 4$ and $\{i, j, k\}$ be a permutation of $\{1, 2, 3\}$. If exactly two elements $a_{1,i+1}, a_{1,j+1}$ are negative, then A is cop $\Leftrightarrow W_1BW_1^T$ and $W_2BW_2^T$ are cop, where

$$W_1 = mat\{e_i, e_j, \overline{V}^{i,k}\}, \quad W_2 = mat\{e_j, \overline{V}^{i,k}, \overline{V}^{j,k}\}.$$

(5) Let $n = 5$ and $\{i,j,k,l\}$ be a permutation of $\{1,2,3,4\}$. If exactly two elements $a_{1,i+1}, a_{1,j+1}$ are negative, then A is cop $\Leftrightarrow P_1 B P_1^T, P_2 B P_2^T$, and $P_3 B P_3^T$ are cop. Here

$$P_1 = mat\{e_i, e_j, \overline{V}^{i,k}, \overline{V}^{i,l}\},$$
$$P_2 = mat\{\overline{V}^{j,l}, e_j, \overline{V}^{i,k}, \overline{V}^{i,l}\},$$

and

$$P_3 = mat\{\overline{V}^{j,l}, e_j, \overline{V}^{i,k}, \overline{V}^{j,k}\}.$$

(6) Let $n = 5$, If exactly three elements, $a_{1,i+1}, a_{1,j+1}, a_{1,k+1}$ are negative, then A is cop $\Leftrightarrow Q_1 B Q_1^T, Q_2 B Q_2^T$, and $Q_3 B Q_3^T$ are all cop. Here

$$Q_1 = mat\{e_i, e_j, e_k, \overline{V}^{i,l}\},$$
$$Q_2 = mat\{e_j, e_k, \overline{V}^{i,l}, \overline{V}^{k,l}\},$$

and

$$Q_3 = mat\{e_j, \overline{V}^{i,l}, \overline{V}^{k,l}, \overline{V}^{j,l}\}.$$

Proof Case 1, 2 are proved in Theorem 4.

Case 3: Here $k = 1$ and $N_{i_1+1} = \{\hat{V}^{i_1,j_l}\}_{l=1}^{n-2}, j_l \neq i_1$, and $vert(T^-) = \{T_{i_1}, N_{i_1+1}\}$. Further, $card(vert(T^-)) = n - 1$, and T^- is a simplex (so no further subdivision is needed). The result follows from Lemma 13 and Lemma 2 together with the observation that the denominators in the expression for $\{\hat{V}^{i_1,j_l}\}_l$ all will be nonzero and have the same sign. Hence $\hat{V}$ is well defined and may be replaced by $\overline{V}$.

Case 4: Here the solid T^- has the vertices $\{T_i, T_j, \hat{V}^{ik}, \hat{V}^{jk}\}$. T^- can be divided into two nonintersecting simplices with vertices $\{T_j, \hat{V}^{jk}, \hat{V}^{ik}\}$ and $\{T_i, T_j, \hat{V}^{ik}\}$. For each such simplex we repeat the arguments of Lemma 13. Arguing as in case 3, it is also seen that $\hat{V}$ is well defined and may be replaced by $\overline{V}$.

Case 5: Here T^- becomes the solid with the six vertices $\{T_i, T_j, \hat{V}^{i,k}, \hat{V}^{i,l}, \hat{V}^{j,k}, \hat{V}^{j,l}\}$, which can be divided into three nonintersecting simplices with vertices $\{T_i, T_j, \hat{V}^{i,k}, \hat{V}^{i,l}\}, \{\hat{V}^{j,l}, T_j, \hat{V}^{i,k}, \hat{V}^{i,l}\}$, and $\{\hat{V}^{j,l}, T_j, \hat{V}^{i,k}, \hat{V}^{j,k}\}$. For each such simplex we repeat the arguments of Lemma 13. Arguing as in case 3, it is also seen that $\hat{V}$ is well defined and may be replaced by $\overline{V}$.

Case 6: Here T^- is the solid with the vertices $\{T_i, T_j, T_k, \hat{V}^{i,l}, \hat{V}^{j,l}, \hat{V}^{k,l}\}$, which can be divided into three nonintersecting simplices with vertices $\{T_i, T_j, T_k, \hat{V}^{i,l}\}, \{T_j, T_k, \hat{V}^{i,l}, \hat{V}^{k,l}\}$, and $\{T_j, \hat{V}^{i,l}, \hat{V}^{k,l}, \hat{V}^{j,l}\}$. Here the reader may consult Figure 5. For each such simplex we repeat the arguments of Lemma 13. Arguing as in case 3, it is also seen that $\hat{V}$ is well defined and may be replaced by $\overline{V}$. □

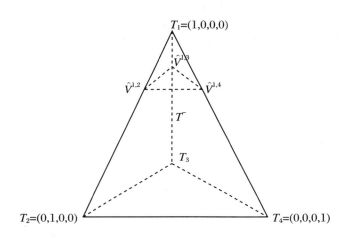

Fig 5 The set T^- for $n=5$ and $a_{13}, a_{14}, a_{15} < 0$

Remark 3.1 *Theorem 14 immediately implies the following way to check whether a given 4×4 matrix is copositive or not. First identify to which of the four cases each row belongs. Then let the row with the lowest case number correspond to $(a_{11}, \overline{a}_1^T)$, and check (17). If (17) is not fulfilled, then A is ncp. Finally, check the appropriate case condition. The number of 3×3 copositivity checks then varies between 1 [(17) does not hold, or (17) holds and case 1 occurs in a row] and 3 [(17) holds, and case 4 occurs in all rows]. A similar analysis can of course be made also for $n=5$.*

We will exemplify Remark 3.1 with a numerical example. Consider a 4×4

symmetric matrix,
$$\begin{pmatrix} 2 & -2 & -1 & 2 \\ -2 & 3 & 2 & -3 \\ 1 & 2 & 1 & 1 \\ 2 & -3 & 1 & 4 \end{pmatrix}.$$

Here the lowest case number is 3. Interchanging the 1st and 4th rows and columns respectively, we obtain

$$A = \begin{pmatrix} 4 & -3 & 1 & 2 \\ -3 & 3 & 2 & -2 \\ 1 & 2 & 1 & -1 \\ 2 & -2 & -1 & 2 \end{pmatrix}.$$

With the notation of Theorem 14, we have

$$a_{11} = 4 > 0, \quad \bar{a}_1^T = (-3, 1, 2),$$

and

$$A_2 = \begin{pmatrix} 3 & 2 & -2 \\ 2 & 1 & -1 \\ -2 & -1 & 2 \end{pmatrix},$$

which can be shown to be cop. Further,

$$B = a_{11}A_2 - \bar{a}_1\bar{a}_1^T = \begin{pmatrix} 3 & 11 & -2 \\ 11 & 3 & -6 \\ -2 & -6 & 4 \end{pmatrix}.$$

This is case 3 with $i_1 = 1$ and $W(1) = mat\{e_1, \overline{V}^{1,2}, \overline{V}^{1,3}\}$, which gives

$$W(1) = \begin{pmatrix} 1 & 0 & 0 \\ a_{13} & -a_{12} & 0 \\ a_{14} & 0 & -a_{12} \end{pmatrix} = \begin{pmatrix} 1 & 0 & 0 \\ 1 & 3 & 0 \\ 2 & 0 & 3 \end{pmatrix},$$

and

$$W(1)BW(1)^T = \begin{pmatrix} 3 & 36 & 0 \\ 36 & 96 & 12 \\ 0 & 12 & 24 \end{pmatrix}$$

is obviously cop. Hence A as well as the original 4×4 matrix is cop.

Remark 3.2 *If $A \in S_n$ with n odd, it is impossible that each row of A contains an odd number of negative off-diagonal elements. Hence e.g. with $A \in S_5$, case 5 (three negative elements) cannot occur in all rows, so this case may be avoided completely by picking another row as $\bar{a}_1$.*

We will end by discussing the cases $n = 2, 3$ using our framework. For $n = 2$ only cases 1 and 2 may occur in Theorem 14, and it is straightforward to verify the equivalence with the conditions (4). For $n = 3$ we will briefly discuss the relation between the conditions in Theorem 14 (only case 1—3 may occur) and Hadeler's conditions (12)—(15). First, the assumptions that $a_{11} \geqslant 0$ and A_2 is copositive imply, by (4), that

$$a_{11} \geqslant 0, \quad \bar{a}_{23} = \sqrt{a_{22}a_{33}} + a_{23} \geqslant 0.$$

Since we may pick any row as $\bar{a}_1$ we may conclude that (12) and (13) hold. For case 1 ($\bar{a}_1 \geqslant 0$) we find, using $\bar{a}_{23} \geqslant 0$, that for $\bar{a}$ in (14)

$$\bar{a} \geqslant a_{13}\sqrt{a_{22}} + a_{12}\sqrt{a_{33}} \geqslant 0,$$

i.e.,(14) is fulfilled.

We now consider case 2. Then B is a 2×2 matrix with elements

$$b_{11} = a_{11}a_{22} - a_{12}^2, \quad b_{12} = a_{11}a_{23} - a_{12}a_{13}, \quad b_{22} = a_{11}a_{33} - a_{13}^2.$$

The copositivity of B implies by (4) that $b_{11}, b_{22} \geqslant 0$, which in turn implies (13) (here we use that a_{12}, a_{13} are negative). Also by (4), $\sqrt{b_{11}b_{22}} + b_{12} \geqslant 0$. For the last inequality we distinguish between the two cases: (i) $b_{12} \geqslant 0$ and (ii) $b_{12} < 0$. For case (i)

$$a_{11}a_{23} + a_{11}\sqrt{a_{22}a_{33}} + a_{13}\sqrt{a_{11}a_{22}} + a_{12}\sqrt{a_{11}a_{33}}$$

$$\geqslant (a_{13} + \sqrt{a_{11}a_{33}})(a_{12} + \sqrt{a_{11}a_{22}}) \geqslant 0,$$

where the last inequality follows from (13), by multiplying with $a_{11}^{-1/2}$ we retrieve (14).

For case (ii), using Lemma 3, we have $b_{11}b_{22} \geqslant b_{12}^2$. Straightforward calculations give (15). Finally we note by Remark 3.2 that cases 3 does not need to be inspected.

The work of the first and the third author was supported by the Swedish Natural Science Research Council under contracts F-FU 8448-306 and F-FU 9443 304 respectively. The work of the second author was supported by the Swedish Institute, the Chinese National Science Foundation, and The Third World Academy of Sciences. The authors also want to thank the referee for constructive criticism.

References

[1] Baumert L D. Extreme Copositive Quadratic Forms II[J]. Pacific J. Math., 1967, 20:1-20.

[2] Berman A. Cones, Matrices, and Mathematical Programming[M]//Econom. and Math. Systems 79, Springer-Verlag, 1973.

[3] Chang G, Sederberg T W. Nonnegative Quadratic Bézier Triangular Patches[J]. Comput. Aided Geom. Design, 1994,11:113-116.

[4] Cottle R W, Habetle G J, Lemke C E. Quadratic Forms Semi-definite over Convex Cones[C]//Princeton Symposium on Mathematical Programming. Princeton, N.J.: Princeton U.P., 1970: 551-565.

[5] Cottle R W, Habetler G J, Lemke C E. On Classes of Copositive Matrices[J]. Linear Algebra Appl., 1970, 3:295-310.

[6] Hadeler K P. On Copositive Matrices[J]. Linear Algebra Appl., 1983, 49:79-89.

[7] Hall Jr. M. Combinatorial Theory[M]. 2nd ed. New York: Wiley, 1986.

[8] Hoffman A J, Pereira F. On Copositive Matrices with -1,0,1 entries[J]. J. Combin. Theory Ser. A, 1973, 14:302-309.

[9] Li P, Feng Y Y. Criteria for Copositive Matrices of Order Four[J]. Linear Algebra Appl. 1993, 194:109-124.

[10] Loewy R, Schneider H. Positive Operators on Then-dimensional Ice-cream Cone[J]. J. Math. Anal. Appl., 1975, 49:375-392.

[11] Martin D H, Jacobson D H. Copositive Matrices and Definiteness of Quadratic Forms Subject to Homogenous Linear Inequality Constraints[J]. Linear Algebra Appl., 1981 (35):227-258.

[12] Micchelli C A, Pinkus A. Some Remarks on Nonnegative Polynomials on Polyhedra[M]//Probability, Statistics and Mathematics (papers in honor of S. Karlin). Boston: Academic Press,1989: 163-185.

[13] Motzkin T S. Copositive Quadratic Forms[R]. National Bureau of Standards Report 1818. 1952: 11-22.

[14] Murty K G, Kabadi S N. Some NP-complete Problems in Quadratic and Nonlinear Programming[J]. Math. Programming, 1987, 39:117-129.

[15] Nadler E. Nonnegativity of Bivariate Quadratic Functions on a Triangle[J]. Comput. Aided Geom. Design, 1992, 9:195-205.

[16] Väliaho H. Criteria for Copositive Matrices[J]. Linear Algebra Appl, 1986, 81: 19-34.

[17] Väliaho H. Testing the Definiteness of Matrices on Polyhedral Cones[J]. Linear Algebra Appl., 1988, 101:135-165.

12

国际数学奥林匹克竞赛

1962～1964 年，北京市举行了高中生数学竞赛，倡导者和组织者是中国科技大学副校长兼数学系主任华罗庚教授. 他以极大的热情，邀集了中科院数学所、北京大学、北京师范大学、中国科技大学的数学工作者，从举办讲座、反复推敲命题，到考试评卷，事无巨细，都由他亲自过问. 参加成员有我校的龚昇先生，当年他是北京数学会的秘书长，还有我校的曾肯成先生，他是命题的主力，专门出些怪题和难题，接下来是科大的年轻助教史济怀、常庚哲，跟着跑龙套.

我记得北京市 1963 年高中三年级的复赛试题，是我提出来的.

例 1 在一个边长为 1 的正方形内任给定 9 个点，试证明：在以这些点为顶点的各三角形中，必有一个三角形，它的面积不大于 $\frac{1}{8}$.

这是我第一次命题并得以发表，现在看来，非常简单和粗浅，用到抽屉原则和三角形的面积.

本题在近 800 页的《Putnam and Beyond》（作者 R. Gelca, T. Andreescu, Springer, 2007）的第 13 页上，作为一个例题，有详细的解答，并指明了来自中国数学竞赛.

可惜好景不长，1965 年北京市高中数学竞赛停办，源于政治气候的转变. 批判"三家村"，"文化大革命"的前奏开始了. 在"文革"中，一代宗师华罗庚因为支持过数学竞赛，受到莫须有的批判——培养封、资、修，复辟资本主义.

1976 年 10 月, 粉碎"四人帮"以后, 科学的春天来到了, 华先生怀着满腔的喜悦, 为中学生和数学老师讲数学竞赛, 在数学杂志上发表文章, 提出大师的见解.

1985 年 6 月 12 日在日本东京, 华罗庚教授不幸与世长辞, 没有看到 1986 年第一届中国数学冬令营, 更没有看到当年第一次中国 6 名选手在国际数学奥林匹克竞赛 (IMO) 取得的好成绩.

1986 年第一届中国数学冬令营在南开大学举办, 其后依次是北京大学 (1987)、复旦大学 (1988) 和中国科技大学 (1989). 我连续参加了四届主试委员会 (或称命题委员会) 的工作, 主要的目的是, 从冬令营选拔优秀学生, 进入国家集训队, 最后确定 6 位学生代表中国参加 IMO.

对于中国数学奥林匹克, 1986 年是值得纪念的年份. 第一件, 第一次以"满队"(6 名选手) 参加 IMO; 第二件, 第一次有中国数学工作者的命题选入 IMO 试题. 我愿意谈谈命题细节.

1986 年 3 月, 我与吉林大学齐东旭教授在浙江大学参加"计算几何"学术研讨会. 由于初等数学是我们的共同爱好, 所以它成了我们晚间聊天的话题. 齐东旭向我谈起 1985 年"五四青年智力竞赛"有这样一道题目:

例 2 地面上有 A, B, C 三点, 一只青蛙位于地面上距 C 点为 27 厘米的 P 点处. 青蛙第一步从 P 点跳到关于 A 点的对称点 P_1 点, 第二步从 P_1 点跳到关于 B 点的对称点 P_2 点, 第三步从 P_2 点跳到关于 C 点的对称点 P_3 点, 第四步从 P_3 点跳到关于 A 点的对称点 P_4 点……按这种方式一直跳下去, 若青蛙在 1985 步跳到了 P_{1985} 点, 问 P 点与 P_{1985} 点相距多少厘米?

解 1985 是一个很大的数目. 如果你想实实在在地一步一步地把青蛙的行踪画出来, 那是不可想象的事, 谁也没有那么多的时间和耐心来重复这种单调无味的操作. 有理由相信, 在青蛙跳的过程中, 一定可以找到某种简单的规律.

从一点出发找出关于某一固定点的对称点, 是由点到点的变换. 为了解这一个题目, 初中数学课本中的"中点公式"就足以够用, 设 (x_1, y_1) 与 (x_2, y_2) 是直角坐标系中的任何两点, 那么连接这两点的线段的中点的坐标由公式

$$\left(\frac{x_1+x_2}{2}, \frac{y_1+y_2}{2}\right)$$

给出.

现在设在某个平面直角坐标系中，$A = (x_a, y_a)$，$B = (x_b, y_b)$，$C = (x_c, y_c)$ 且 $P = (x_0, y_0)$，而 $P_i = (x_i, y_i)$，$i = 1, 2, \cdots, 1985$. 我们只需关注横坐标的变化.

由于 A 点是 P 点与 P_1 点的中点，依中点公式得

$$x_a = \frac{x_0 + x_1}{2},$$

即

$$x_1 = 2x_a - x_0.$$

由同样的理由

$$x_2 = 2x_b - x_1 = 2x_b - 2x_a + x_0,$$
$$x_3 = 2x_c - x_2 = 2x_c - 2x_b + 2x_a - x_0,$$

令 $k = 2(x_c - x_b + x_a)$，这是一个定数，所以

$$x_3 = k - x_0.$$

注意到青蛙由 P_3 点跳到 P_6 点的过程与它从 P 点跳到 P_3 点的过程完全一样，因此利用上式可知

$$x_6 = k - x_3 = k - (k - x_0) = x_0.$$

同理 $y_6 = y_0$，即 $P_6 = P$. 这说明，青蛙跳过六次之后就回到了原来的出发点. 用数学的语言来说，这种变换具有周期性，周期为 6，也就是说

$$P = P_6 = P_{12} = P_{18} = P_{24} = \cdots,$$

由于 $1985 = 6 \times 330 + 5$，故

$$P_{1985} = P_5,$$

而 P_5 点与 $P_6(= P)$ 点是关于 C 点对称的两点，所以

$$\overline{P_5 P} = 2\overline{CP} = 2 \times 27 = 54 \text{（厘米）},$$

这就是我们的答案. **证毕**.

齐东旭问：有没有再推广的可能性？

青蛙的"对称跳"的特点是：走直线，不拐弯，就像大家玩的最普通的"跳棋"一样. 如果设想青蛙有更高的智商，会拐弯，当会出现另一番景象.

平面上任意给定不同的 3 点 A, B, C. 设平面中有一点 P_0，从 P_0 连一直线到达 A，设 θ 为某一个角，青蛙到达 A 点之后，左拐 θ，沿直线使 $\overline{P_0A} = \overline{AP_1}$，青蛙到达后，从 P_1 沿着直线到达 B，左拐 θ 沿着直线使得 $\overline{P_1B} = \overline{BP_2}$ 到达 P_2，青蛙从 P_2 走直线到 C 后，又左拐 θ 沿直线使 $\overline{P_2C} = \overline{CP_3}$ 到达 P_3. 再从 P_3 沿直线到达 A……周而复始，继续下去. 有什么结果？会不会有周期现象？

对我们来说，我们的题目用复数表示最为简洁而方便. "左转弯 θ 运动"就代表 $e^{i\theta} = \cos\theta + i\sin\theta$:

$$(A - P_0)e^{i\theta} = P_1 - A.$$

为了记号简单，令 $u = e^{i\theta}$，我们便有

$$P_1 = (1+u)A - uP_0.$$

用同样的方法可得以下一系列的等式：

$$P_2 = (1+u)B - uP_1$$
$$= (1+u)(B - uA) + u^2 P_0,$$

进一步有 $P_3 = (1+u)C - uP_2$，也就是

$$P_3 = (1+u)(C - uB + u^2 A) - u^3 P_0.$$

由于从 P_3 到 P_6 与从 P_0 到 P_3 的过程完全一样，利用上式可得

$$P_6 = (1+u)(C - uB + u^2 A) - u^3 P_3.$$

最后得到

$$P_6 = (1-u^3)(1+u)(C - uB + u^2 A) + u^6 P_0.$$

现在令 $u^6 = 1$，可得

$$P_6 = (1-u^3)(1+u)(C - uB + u^2 A) + P_0,$$

如果 $P_6 = P_0$，便有

$$(1-u^3)(1+u)(C - uB + u^2 A) = 0.$$

若 $u^3-1=0$ 或者 $u+1=0$，上面那个方程会自动被满足，不管 3 点 A, B, C 如何选取，相应的 4 个根是 $u=1, \mathrm{e}^{2\pi\mathrm{i}/3}, \mathrm{e}^{-2\pi\mathrm{i}/3}, -1$，也就是 $\theta=0, \dfrac{2\pi}{3}, -\dfrac{2\pi}{3}, \pi$. 6 次单位根中另外 2 个是 $u=\mathrm{e}^{\pi\mathrm{i}/3}$ 和 $\mathrm{e}^{-\pi\mathrm{i}/3}$，在这些场合中，方程

$$C - uB + u^2 A = 0$$

成立. 因为 $u^2 = u - 1$，方程变为

$$C - A = (B - A)u,$$

这里 θ 是 $\dfrac{\pi}{3}$ 或 $-\dfrac{\pi}{3}$. 也就是说，当 $\theta = \dfrac{\pi}{3}$ 时 ABC 是正向的等边三角形（图 12.1），而当 $\theta = -\dfrac{\pi}{3}$ 时 ABC 是负向的等边三角形（图 12.2）. 因 1986 是 6 的整倍数. 两个结论中的第一个结论，正是 IMO 选手们所需要的.

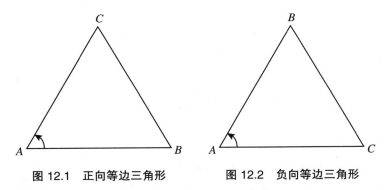

图 12.1　正向等边三角形　　　图 12.2　负向等边三角形

现在，我们撇开"对称跳""青蛙"之类的形象语言，直接进入 IMO 的数学题目.

例 3　在平面上有一个三角形 $A_1 A_2 A_3$ 和一个给定的点 P_0，定义 $A_s = A_{s-3}$，对于所有的 $s \geqslant 4$. 构造点列 $P_1, P_2, P_3, \cdots$，使得以 A_{k+1} 为中心，从 P_k 沿着顺时针方向旋转 $120°$，并记为 P_{k+1}，这里 $k=0,1,2,\cdots$. 证明：如果 $P_{1986} = P_0$，那么三角形 $A_1 A_2 A_3$ 是等边三角形.

除了上面的复数证明之外，还有其他两个证明供读者参考，这一本书叫做《International Mathematical Olympiads, 1986~1999》，美国数学协会出版，作者是波兰数学家 Marcin E. Kuczma.

证法 1　考察平面上的变换，由复合数定义：$f = r_3 \circ r_2 \circ r_1$，这里 r_j 是关于 A_j 的顺时针方向旋转 $120°$，映射 f 保持长度和每一个向量的方向，因此，它就是一些向量的平移. 根据问题的条件，$f(P_0) = P_3$，而由周期

性，$f^n(P_0) = P_{3n}$，这里符号 f^n 表示 n 重复合（迭代）$f \circ f \circ \cdots \circ f$. 因为 f 是 v 的平移，f^n 是向量 nv 的平移. 因 $n = 662$，我们有 $P_{3n} = P_{1986} = P_0$. 因此 v 是零向量，意味着 f 是恒等映射. 记点是 B，于是 $A_1 = f(A_1) = r_3(r_2(r_1(A_1))) = r_3(r_2(A_1)) = r_3(B)$.

两个等腰三角形 A_1A_2B 和 BA_3A_1 不是重合的，它们有相等的角（$\angle A_2 = \angle A_3 = 120°$）并且有公共的底边 A_1B，因而它们是全等的. 于是 $A_1A_2BA_3$ 是一菱形（两个角是 60° 和 120°）（图 12.3），这表明 $A_1A_2A_3$ 是一个等边三角形. **证毕**.

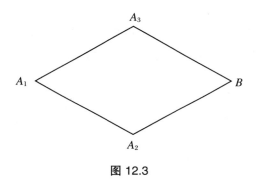

图 12.3

证法 2　仍用复数解法.

给定的点 $A_k, P_k, k = 0, 1, 2, \cdots$ 代表着复数 a_k, p_k. 令记号 r_1, r_2, r_3 和 f 表示与证法 1 中相同的映射. 旋转 r_k 表示从任何一点（复数）z 变到它的像

$$r_k(z) = (z - a_k)\lambda + a_k,$$

这里 λ 是三次单位根

$$\begin{aligned}\lambda &= \cos\left(\frac{2}{3}\right)\pi - \mathrm{i}\sin\left(\frac{2}{3}\right)\pi \\ &= -\frac{1}{2}(1 + \sqrt{3}\mathrm{i}).\end{aligned}$$

因为 $\lambda^3 = 1$，所以

$$\begin{aligned}f(z) &= \{[(z - a_1)\lambda + a_1 - a_2]\lambda + a_2 - a_3\}\lambda + a_3 \\ &= (z - a_1)\lambda^3 + (a_1 - a_3)\lambda^2 + (a_2 - a_3)\lambda + a_3 \\ &= z + w,\end{aligned}$$

这里
$$w = (a_1 - a_2)\lambda^2 + (a_2 - a_3)\lambda + (a_3 - a_1)$$
$$= (\lambda - 1)[(a_1 - a_2)(\lambda + 1) + (a_2 - a_3)].$$

根据假设，$p_0 = p_{1986} = f^{662}(p_0) = p_0 + 662w$，故 $w = 0$. 因为 $\lambda - 1 \neq 0$，我们得到
$$(a_1 - a_2)(\lambda + 1) + (a_2 - a_3) = 0.$$

注意到 $a_1 - a_2 \neq 0$（点 a_1, a_2, a_3 是一个三角形的三个顶点），用 $a_1 - a_2$ 相除，得到
$$\frac{a_3 - a_2}{a_1 - a_2} = \lambda + 1 = \frac{1 - \sqrt{3}\mathrm{i}}{2}$$
$$= \cos\left(\frac{\pi}{3}\right) - \mathrm{i}\sin\left(\frac{\pi}{3}\right).$$

这个方程的几何意义是：$A_1 A_2 A_3$ 是等边三角形，而且是负向的.

1987 年元月，中国数学冬令营在北京大学举行. 命题的背景居然与我们的 Bézier 网有关.

这个命题表述如下：

例 4 把一个给定的等边三角形 ABC 的各边都分成 n 等份，过各分点作平行于其他两边的直线，将这个等边三角形分成 n^2 个小等边三角形 (图 12.4)，这些小三角形的每一个顶点叫做**结点**，在每一个结点上放置了一个实数. 已知：

(1) A, B, C 三个顶点上放置的实数分别是 a, b, c；

(2) 在每两个有公共边长的小三角形组成的菱形中，两组相对顶点上放置的实数之和相等.

试求所有的结点上的总和 S_n.

我们不忙着来证明这个定理. 先考察有下列关系的 5 个结点 (图 12.5).

由最重要的性质，我们有两个等式
$$y + u = x + v,$$
$$y + v = z + u,$$

两式相加并化简得 $2y = x + z$. 而这个等式又可写成 $y - x = z - y$，这表明，**任一共线的结点所放的数成一等差数列，不管等差数列多长**.

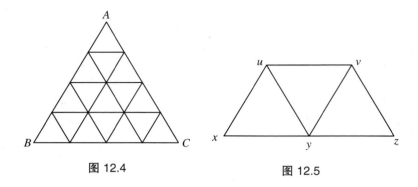

图 12.4　　　　　　　图 12.5

这种特殊的数据表明，Bézier 网是共面的，这是既充分又必要条件.

我们来看特殊的情况. 令 $a=1$, $b=c=0$，Bézier 网的数据如图 12.6 所示.

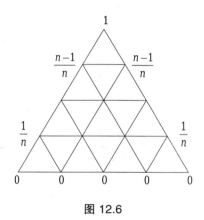

图 12.6

共面的数据是

$$1 \cdot \left(\frac{n}{n}\right) + 2 \cdot \left(\frac{n-1}{n}\right) + 3 \cdot \left(\frac{n-2}{n}\right) + \cdots + (n-1) \cdot \left(\frac{2}{n}\right) + n \cdot \left(\frac{1}{n}\right)$$

$$= \sum_{k=1}^{n} \frac{k}{n}(n-k+1)$$

$$= \left(\frac{n+1}{n}\right) \sum_{k=1}^{n} k - \frac{1}{n} \sum_{k=1}^{n} k^2$$

$$= \frac{1}{6}\left[3(n+1)^2 - (n+1)(2n+1)\right]$$

$$= \frac{1}{6}(n+1)(n+2).$$

最后的总和是

$$S_n = \frac{(n+1)(n+2)}{6}(a+b+c).$$

我造出这个命题,是根据相关的数据成等差数列,来推断 Bézier 网是共面的. 不知道 Bézier 网的学生大可放心,只要知道等差数列的一般知识,就能解决问题,用不同长度的数列,多算几次就可以了. 不过一定要小心,计算容易出错.

令人惊异的是,1986 年冬令营的一个选手,上海向明中学的学生潘子刚(他后来成了我国参加第 28 届国际中学生奥林匹克的代表队员)所给出的解答充分地利用了对称性,为此,他获得了该届冬令营的特别奖. 令我们感叹:数学是年轻人的!

潘子刚的解答 将同一个等边三角形按不同的方位放置,即按逆时针方向旋转 120°. 三个图形叠加起来,得到一张等边三角形数表,三个顶点上的数都是 $a+b+c$(图 12.7).

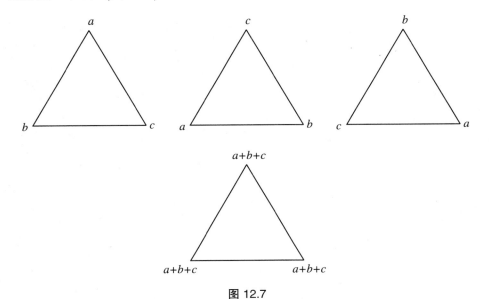

图 12.7

由于原来的三张数表有性质 (2),所以三个数表叠加起来,也有性质 (2),

由此可知，每一个结点上所放置的数都是 $a+b+c$，因此其总和为
$$\frac{(n+1)(n+2)}{2}(a+b+c).$$
但因它是由具有同一和数 S_n 的三张三角形数表叠加起来的，故上面的数为 $3S_n$，所以
$$S_n = \frac{(n+1)(n+2)}{6}(a+b+c).$$
比我的做法要简单得多.

13

Whitney 定理

1984 年，我在《美国数学月刊》上发表一个初等题目征解，翻译成中文就是：设函数 f 直到 $n+1$ 阶导数有定义，在 $x=0$ 的邻域连续，满足 $f(0)=0$，求证

$$\lim_{x\to 0}\frac{\mathrm{d}^k}{\mathrm{d}x^k}\left(\frac{f(x)}{x}\right)=\frac{1}{k+1}f^{(k+1)}(0), \tag{1}$$

这里 $k=0,1,\cdots,n$.

天天等着解答，等来的竟然是坏的事情：1965 年，也是《美国数学月刊》，544 页，上面的内容跟我发表的题目一模一样！相隔不到 20 年. 我甚至埋怨审查人粗枝大叶，不负责任.《美国数学月刊》是百年老字号，世界上凡是有大学数学系的地方，大都有《美国数学月刊》. 问题征解的栏目，是我最喜欢看的. 现在让我如何面对编辑和读者？会不会上"黑名单"？会不会说我偷窃别人的成果？

后来转念一想，如果故意偷窃他人的成果，还在同一个杂志"作案"，那这不是自投罗网吗？美国的同行不计较我的疏忽，让我非常感动. 我在西方数学杂志发表过很多文章，至少有 7 种杂志有我提出的"问题栏目"，都对我平等对待，没有另眼相看.

其实，1965 年的命题，也不是新的. 在 1966 年公布解答的时候，某

教授已经指出，它是属于美国大名鼎鼎的数学家 Hassler Whitney 教授的. Whitney 定理单变量函数部分是出现在《Advanced Calculus：An Introduction to Analysis》(Watson Fulks, Wiley, 1961: 140) 上的一个问题. Whitney 将其推广到多变量函数的情形.

　　H. Whitney（1907~1989），专长微分几何，早年研究过图论，是 1982 年沃尔夫数学奖得主.

Problems and Solutions

1   Elementary problems

E3031. *Proposed by Gengzhe Chang, University of Science and Technology of China, Hefei, Anhui, China.*

Suppose function f together with its derivatives of order up to $n+1$ is defined and continuous in a neighborhood of $x=0$ and satisfies $f(0)=0$. Prove that
$$\lim_{x \to 0} \frac{d^k}{dx^k}\left[\frac{f(x)}{x}\right] = \frac{1}{k+1} f^{(k+1)}(0)$$
for $k = 0, 1, 2, \cdots, n$.

Solution by A. Villani, Seminario Matematico, Catania, Italy. By Leibniz's Formula, we have for $k = 0, 1, 2, \cdots, n$
$$\left(\frac{d}{dx}\right)^k \left(f(x) \cdot \frac{1}{x}\right) = \frac{\psi(x)}{x^{k+1}},$$
where
$$\psi(x) = \sum_{r=0}^{k} \binom{k}{r} (-1)^r r! \, x^{k-r} f^{(k-r)}(x).$$
Since $\lim_{x \to 0} \psi(x) = 0$, we may find the limit of the quotient as $x \to 0$ by L'Hôpital's Rule. The sum $\psi'(x)$ telescopes to $x^k f^{(k+1)}(x)$, so
$$\lim_{x \to 0} \left(\frac{d}{dx}\right)^k \left(f(x) \cdot \frac{1}{x}\right) = \lim_{x \to 0} \frac{f^{(k+1)}(x)}{k+1} = \frac{f^{(k+1)}(0)}{k+1}.$$

Alternative solution by Hong Oh Kim, Kyungpook National University, Korea.
Since $f(0) = 0$, we can write
$$\frac{f(x)}{x} = \int_0^1 f'(xt)\,dt.$$
Differentiation under the integral sign is justified, so
$$\left(\frac{d}{dx}\right)^k \left(\frac{f(x)}{x}\right) = \int_0^1 f^{(k+1)}(xt) t^k \, dt.$$

Taking the limit under the integral sign is also justified, so

$$\lim_{x\to 0}\left(\frac{\mathrm{d}}{\mathrm{d}x}\right)^k \frac{f(x)}{x} = \int_0^1 f^{(k+1)}(0)t^k \mathrm{d}t = \frac{f^{(k+1)}(0)}{k+1}.$$

A. E. McKaig observed that this problem has appeared in the Monthly once before, viz, E 1789 [1965, 544]. Its solutions were discussed in [1966, 779].

An additional reference was provided by W. Gautschi. The numerical stability of the recursion

$$xr_k(x) + kr_{k-1}(x) = f^{(k)}(x),$$

where

$$r_k(x) = \left(\frac{\mathrm{d}}{\mathrm{d}k}\right)^k \frac{f(x)}{x},$$

is analyzed in W. Gautschi, Computation of Successive Derivatives of $\frac{f(z)}{z}$, Math Comp, 1966, 20: 209-214. See also W. Gautschi, Zur Numerik rekurrenter Relation, Computing, 1972, 9: 107-125, in particular Example 5.2.

Also solved by 48 other readers.

2 A special case of a theorem of H. Whitney

E 1789 [1965, 544]. Proposed by R. A. Bell, Kansas City, Mo. Suppose that $g(x)$ has its first $n+1$ derivatives defined and continuous in $[-1,1]$. Define $y(x) = g(x)/x$ for $x \neq 0$ and $y(0) = g'(0)$. If $g(0) = 0$, prove that $y^{(n)}(0) = d^n y/dx^n|_{x=0}$ exists and equals $g^{(n+1)}(0)/(n+1)$.

I. *Solution by W. O. Egerland, USA Nuclear Defense Laboratory, Edgewood Arsenal.* For $x \neq 0$ we have

$$xy(x) = g(x) = \int_0^x g'(t)\mathrm{d}t \quad \text{or} \quad y(x) = \int_0^1 g'(xu)\mathrm{d}u.$$

Since $g'(xu)$ is $(n-1)$ times continuously differentiable with respect to x in the rectangles $R_1[0 \leq u \leq 1, 0 < x_1 \leq x \leq 1]$ and $R_2[0 \leq u \leq 1, -1 \leq x \leq x_2 < 0]$, we obtain by Leibniz's Rule

$$y^{(n-1)}(x) = \int_0^1 u^{n-1} g^{(n)}(xu) du, \quad x \neq 0,$$

so that

$$\frac{1}{x}\left[y^{(n-1)}(x) - \frac{g^{(n)}(0)}{n}\right] = \frac{1}{x^{n+1}} \int_0^x t^{n-1}[g^{(n)}(t) - g^{(n)}(0)] dt$$
$$\to \frac{g^{(n+1)}(0)}{n+1}$$
$$= y^{(n)}(0).$$

as $x \to 0^+$ or $x \to 0^-$, by L'Hôpital's Rule.

The proof shows that only the existence of $g^{(n+1)}(x)$ at $x = 0$ is needed to establish the assertion of the problem.

II. *Solution by M. S. Klamkin, Mathematical and Theoretical Sciences Scientific Laboratory, Ford Motor Company, Dearborn, Michigan.* Let

$$D^m[g(x)/x] = F_m(x)/x^{m+1},$$

so that $DF_0(x) = D\{xD^0[g(x)/x]\} = Dg(x) = x^0 Dg(x)$. Assume that

$$DF_m(x) = x^m D^{m+1} g(x). \tag{2}$$

Then

$$\frac{F_{m+1}(x)}{x^{m+2}} = D^{m+1}\frac{g(x)}{x} = D\frac{F_m(x)}{x^{m+1}} = \frac{xDF_m(x) - (m+1)F_m(x)}{x^{m+2}},$$

whence $F_{m+1}(x) = xDF_m(x) - (m+1)F_m(x)$ and

$$DF_{m+1}(x) = xD^2 F_m(x) - (m+1)DF_m(x)$$
$$= xD[x^m D^{m+1} g(x)] - (m+1)x^m D^{m+1} g(x)$$
$$= x^{m+1} D^{m+2} g(x).$$

It follows by induction that (2) is true for all nonnegative integers m.

By L'Hospital's Rule, $\lim_{x\to 0} D^n[g(x)/x]$ will exist if $\lim_{x\to 0}[DF_n(x)/Dx^{n+l}]$ exists. By (2), this latter limit is $g^{(n)}(0)/(n+1)$.

III. *Comment by L. E. Pursell, Grinnell College.* This problem is a special case of a theorem proved by Hassler Whitney, *Differentiability of the remainder term in Taylor's formula*, Duke Math. J., 1943, 10: 153-158, concerning the derivatives of the "remainder quotients", $\dfrac{R_{n-1}(x)}{x^n}$, where

$$R_{n-1}(x) = g(x) - \dfrac{\sum_{k=0}^{n-1} g^{(k)}(0)x^k}{k}!.$$

Professor Whitney also extends his results to functions of several variables. Whitney's theorem for functions of one variable also appears as a problem in Watson Fulks, *Advanced Calculus: An Introduction to Analysis*, Wiley, 1961, p. 140, problem C1.

I devised a proof similar to that in Solution I above several years ago after encountering the integral transform $\int_0^1 g'(xt)\mathrm{d}t$ in K. Nomizu, *Lie Groups and Differential Geometry*, Math. Soc. of Japan, 1956, p. 7. A similar proof for a slightly different theorem appears in Sigurdur Helgason, *Differential Geometry and Symmetric Spaces*, Academic Press, 1962, pp. 9-10.

Also solved by E. O. Buchman, J. L. Gieser, Richard Gisselquist, D. M. Goldschmidt, R. W. Hansell, Stephen Hoffman, S. C. King, E. S. Langford, J. C. Lazzara, D. C. B. Marsh, Y. Mayerben-David, Norman Miller, J. M. Perry, Simeon Reich (Israel), Al Somayajulu, Sidney Spital, H. H. Wong, and the proposer. Partial solutions by P. J. Campbell, G. A. Fisher, D. M. Hancasky, J. C. Hickman, D. R. Lehman, N. T. Sheth, R. Sivaramakrishnan (India), K. L. Yocom, and David Zeitlin.

Some of those listed as partial solvers obtained $y^{(k)}(x)$ by differentiating Taylor's Formula throughout but forgot that the c in the remainder $\dfrac{x^{n+1}g^{(n+1)}(c)}{(n+1)!}$ is also a function of x. Others assumed that $xy^{(n+l)}(x)$ has limit 0 as $x \to 0$.

14

命题五则

西方数学有影响的杂志上,大都设有"问题与解答"的栏目,例如《美国数学月刊》《数学杂志》《SIAM Review》等等. 我的刊登出来的问题,起码有 20 多道. 我挑选了五题,其中两题从英文翻译成中文,另外三题原文(英文)照登.

1 命题一

设函数 f 在 $[0,1]$ 上连续,在 $(0,1)$ 内可导,并且 $f(0)=0, f(1)=1$. 又设 $k_1, k_2, \cdots, k_n$ 是任意的 n 个正数. 求证:在 $(0,1)$ 中存在 n 个互不相同的数 $t_1, t_2, \cdots, t_n$,使得

$$\sum_{i=1}^{n} \frac{k_i}{f'(t_i)} = \sum_{i=1}^{n} k_i.$$

(选自《Mathematics Magazine》,1981 年,54 卷第 140 页.)

证明 令 $K = k_1 + k_2 + \cdots + k_n, y_0 = 0$ 以及

$$y_i = \frac{1}{K}(k_1 + \cdots + k_i), \quad i = 1, 2, \cdots, n.$$

所以得到 $0 = y_0 < y_1 < \cdots < y_n = 1$. 取 $x_0 = 0, x_n = 1$. 在 $[0,1]$ 上对连续函数 f 用介值定理,可以求得一点 $x_1 \in (0,1)$ 使 $f(x_1) = y_1$,再在 $[x_1, 1]$ 上用介值定

理求得一点 $x_2 \in (x_1, 1)$ 使得 $f(x_2) = y_2$. 仿此前进，求出 $x_3 < \cdots < x_{n-1} < 1$, 使得 $f(x_i) = y_i, i = 3, \cdots, n-1$. 总之，我们有 $f(x_i) = y_i, i = 0, 1, 2, \cdots, n$, 在每一个区间 $[x_{i-1}, x_i]$ 上，使用中值定理，求得 $t_i \in (x_{i-1}, x_i)$ 满足

$$y_i - y_{i-1} = f(x_i) - f(x_{i-1}) = f'(t_i)(x_i - x_{i-1}),$$

由此得出

$$\frac{y_i - y_{i-1}}{f'(t_i)} = x_i - x_{i-1},$$

也就是

$$\frac{k_i}{f'(t_i)} = K(x_i - x_{i-1}).$$

将上式对 i 从 1 到 n 求和，得到

$$\sum_{i=1}^{n} \frac{k_i}{f'(t_i)} = K \sum_{i=1}^{n} (x_i - x_{i-1}) = K(x_n - x_0) = K.$$

证毕.

2　命题二

设 n 是自然数，令

$$S_n = 1 + \frac{n-1}{n+2} + \frac{n-1}{n+2} \cdot \frac{n-2}{n+3} + \cdots + \frac{n-1}{n+2} \cdot \frac{n-2}{n+3} \cdots \cdot \frac{1}{2n},$$

求证：

$$\lim_{n \to \infty} \frac{S_n}{\sqrt{n}} = \frac{\sqrt{\pi}}{2}.$$

（选自《The American Mathematical Monthly》，1982 年，89 卷 65 页.）

证明　S_n 可写为

$$S_n = 1 + \sum_{k=1}^{n-1} \frac{(n-1)\cdots(n-k)}{(n+2)\cdots(n+k+1)}$$

$$= 1 + \frac{1}{\binom{2n}{n-1}} \sum_{k=1}^{n-1} \binom{2n}{n+k+1}$$

$$= 1 + \frac{1}{\binom{2n}{n-1}} \left[\binom{2n}{n+2} + \binom{2n}{n+3} + \cdots + \binom{2n}{2n} \right]$$

$$= 1 + \frac{1}{\binom{2n}{n-1}} \left[\sum_{k=0}^{2n} \binom{2n}{k} - 2\binom{2n}{n-1} - \binom{2n}{n} \right] \frac{1}{2}$$

$$= \frac{1}{2} \frac{1}{\binom{2n}{n-1}} 2^{2n} - \frac{1}{2}\left(1 + \frac{1}{n}\right).$$

由此可知

$$\lim_{n \to \infty} \frac{S_n}{\sqrt{n}} = \frac{1}{2} \lim_{n \to \infty} \frac{2^{2n}}{\sqrt{n}\binom{2n}{n-1}}.$$

由 Stirling 公式，得到

$$n! \sim \sqrt{2n\pi}\left(\frac{n}{e}\right)^n, \quad n \to \infty,$$

最后得出

$$\lim_{n \to \infty} \frac{S_n}{\sqrt{n}} = \frac{\sqrt{\pi}}{2}.$$

证毕.

3 命题三

Problem 82-19: A binomial coefficient summation, by Gengzhe Chang (University of Utah).

Show that

$$\sum_{i=k}^{n} \binom{i}{k}\binom{2i}{i}\binom{2n-2i}{n-i} = 4^{n-k}\binom{2k}{k}\binom{n}{k}. \tag{1}$$

Solution by J. Roppert (University of Economics, Vienna, Austria).

Using

$$\binom{-\frac{1}{2}}{n} = (-4)^{-n}\binom{2n}{n},$$

(1) is equivalent to

$$\sum_{i=k}^{n}\binom{i}{k}\binom{-\frac{1}{2}}{i}\binom{-\frac{1}{2}}{n-i}=(-1)^{n+k}\binom{-\frac{1}{2}}{k}\binom{n}{k}, \quad (2)$$

which follows more generally from

$$\sum_{i=k}^{n}\binom{i}{k}\binom{a}{i}\binom{b}{n-i}=\binom{a}{k}\binom{a+b-k}{n-k}, \quad a, b \text{ real.} \quad (3)$$

Equation (3) follows by expanding the summand and rewriting it as the well-known Vandermonde convolution

$$\sum_{j=0}^{n}\binom{a-k}{j}\binom{b}{n-j}=\binom{a+b-k}{n}$$

Editorial note. A. Sidi(The Technion, Haifa, Israel), in his solution, also notes more generally that

$$\sum_{i=k}^{n-l}\binom{i}{k}\binom{n-i}{l}\binom{2i}{i}\binom{2n-2i}{n-i}=4^{n-k-l}\binom{2k}{k}\binom{2l}{l}\binom{n}{k+l} \quad (4)$$

and that (4) will follow from

$$\sum_{i=p}^{n-p}\binom{i}{p}\binom{n-i}{q}\binom{a}{i}\binom{b}{n-i}=\binom{a}{p}\binom{b}{q}\binom{a+b-p-q}{n-p-q} \quad (5)$$

by using (2). Equation (5) follows in a similar way to (3).

O. G. Ruehr (Michigan Technological University) obtains the more general result

$$\sum_{i=k}^{n}\binom{i}{k}\binom{2i}{i}\binom{2n-2i}{n-i}x^{n-i}$$
$$=\binom{n}{k}\sum_{s=k}^{n}\binom{n-k}{s-k}\binom{2s}{s}(4x)^{n-s}(1-x)^{s-k}.$$

G. E. Andrews(Pennsylvania State University) included the following remarks with his solution (similar remarks were also made by R. Askey (University of Wisconsin) in his solution):

It should be pointed out that most problems like this are completely routine exercises if one follows the straightforward algorithms set down in §5 of SIAM Rev., 1974, 16: 441-484. Following the method described there will either lead directly to the desired proof (99.44% of cases) or to a brand new summation of a hypergeometric series (0.56% of cases), an event of some significance.

Although Askey felt that the problem here should not have been published because of the above general method, it should be noted that the solutions given here are considerably shorter than the ones using the above algorithm and this will usually be the case with general methods. Additionally, a number of readers will now be aware of the above general technique and also, more than occasionally, they will come up with more general results than that given in the stated problem. [M.S.K.]

Also solved by L. C. Becker (Christian Brothers College), D. A. Carlson (University of Massachusetts), B. Ganapol (University of Arizona), E. Hansen (Lock heed Missiles & Space Company), I. Lie (Norwegian Defence Research Establishment, Kjeller, Norway), A. A. Jagers (Technische Hogeschool Twente, Enschede, the Netherlands), W.B. Jordan (Scotia, NY), I. I. Kotlarski (Oklahoma State University), O. P. Lossers (2 solutions) (Eindhoven University of Technology, Eindhoven, the Netherlands), H. Prodinger (Vienna Technical University, Vienna, Austria), J. A. Wilson(Iowa State University), P. Y. Wu (National Chiao Tung University, Hsinchu, Taiwan) *and the proposer.*

4 命题四

Problem 83-2: Limit of a power of a matrix, by Gengzhe Chang(Brown University).

Evaluate $\lim_{k \to \infty} T^k$ where T is an $n \times n$ matrix whose (i,j) entry is $J_{n,i}(j/n)$, where

$$J_{n,i}(x) = \binom{n}{i} x^i (1-x)^{n-i}, \qquad i = 1, 2, \cdots, n.$$

The problem is related to the limiting behavior of iterates of the Bernstein polynomials of a fixed function.

Solution by Michael Renardy (Mathematics Research Center, University of Wisconsin-Madison).

We have $T_{nn} = 1$, and $T_{in} = 0$ for $i < n$, hence T has an eigenvalue 1. Moreover, for $j < n$, we have

$$\sum_{i=1}^{n-1} |T_{ij}| = \sum_{i=1}^{n-1} \binom{n}{i} \left(\frac{j}{n}\right)^i \left(1 - \frac{j}{n}\right)^{n-i} < \sum_{i=0}^{n} \binom{n}{i} \left(\frac{j}{n}\right)^i \left(1 - \frac{j}{n}\right)^{n-i} = 1,$$

hence the remaining eigenvalue of T have modulus less than 1. The right-hand eigenvector belonging to eigenvalue 1 is $V = (0, 0, \cdots, 0, 1)^{\mathrm{T}}$, and the left eigenvector is $(\frac{1}{n}, \frac{2}{n}, \cdots, 1) = W$. Here we have used

$$\sum_{i=1}^{n} \binom{n}{i} \left(\frac{j}{n}\right)^i \left(1 - \frac{j}{n}\right)^{n-i} \frac{i}{n} = \sum_{i=1}^{n} \binom{n-1}{i-1} \left(\frac{j}{n}\right)^i \left(1 - \frac{j}{n}\right)^{n-i}$$

$$= \frac{j}{n} \sum_{i=0}^{n-1} \binom{n-1}{i} \left(\frac{j}{n}\right)^i \left(1 - \frac{j}{n}\right)^{n-1-j} = \frac{j}{n}.$$

The limit $\lim_{k \to \infty} T^k$ is the normalized dyadic product of the two eigenvectors:

$$\lim_{k \to \infty} T^k = \frac{vw}{v \cdot w} = \begin{pmatrix} 0 & 0 & \cdots & 0 \\ 0 & 0 & \cdots & 0 \\ \vdots & \vdots & & \vdots \\ \frac{1}{n} & \frac{2}{n} & \cdots & 1 \end{pmatrix}.$$

Also solved by A. A. Jagers (Technishce Hogeschool Twente, Enschede, the Netherlands), W. B. Jordan (Scotia, NY) and E. L. Wachspress (University of Tennessee), Ingemar Kinnmark (Princeton University), O. P. Lossers (Eindhoven University of Technology, the Netherlands), A. Sidi (The Technion, Haifa, Israel) and H. Van Haeringen (Delft University of Technology, the Netherlands), ***and the proposer.***

5 命题五

Problem 84-15: A system of ordinary differential equations, by Gengzhe Chang (University of Science and Technology of China, Hefei, Anhui, China).

Find the solution of the system of ordinary differential equations

$$tX'(t) = AX(t), \text{ with } X(1) \text{ given},$$

where

$$X(t) = \begin{bmatrix} x_1(t) \\ \vdots \\ x_n(t) \end{bmatrix}$$

and A is an $n \times n$ constant matrix with eigenvalues $1, 2, \cdots, n$.

Solution by John A. Crow (Student, California State University, Fullerton).

Consider the general system of ordinary differential equations

$$X'(t) - f'(t)AX(t) = 0, \tag{6}$$

where X is a column vector and A is an $n \times n$ constant matrix with distinct nonzero eigenvalues $\lambda_1, \cdots, \lambda_n$. As may be verified by direct substitution, the general solution is

$$X(t) = \exp\{Af(t)\} \cdot \alpha, \tag{7}$$

where α is an arbitrary constant column vector. Now suppose $X(t_0)$ is specified. Then using the fact that $\{\exp(At)\}^{-1} = \exp(-At)$, it follows that

$$X(t) = \exp\{A[f(t) - f(t_0)]\} \cdot X(t_0). \tag{8}$$

Without loss of generality, assume $f(t_0) = 0$. Since the eigenvalues are distinct and nonzero, then there is a similarity transformation P such that $B = P^{-1}AP = \text{diag}\{\lambda_1, \cdots, \lambda_n\}$. Using the relation $\exp(At) = P\exp(Bt)P^{-1}$, and the fact that

$$\exp(Bt) = \begin{bmatrix} e^{\lambda_1 t} & & 0 \\ & \ddots & \\ 0 & & e^{\lambda_n t} \end{bmatrix},$$

it follows that

$$X(t) = P \begin{bmatrix} e^{\lambda_1 f(t)} & & 0 \\ & \ddots & \\ 0 & & e^{\lambda_n f(t)} \end{bmatrix} P^{-1} X(t_0). \tag{9}$$

In the special case where $f'(t) = 1/t, t_0 = 1$, and $\lambda_j = j$, then (9) reduces to

$$X(t) = P \begin{bmatrix} t^1 & & 0 \\ & t^2 & \\ & & \ddots \\ 0 & & & t^n \end{bmatrix} P^{-1} X(1).$$

These results can be extended to nonhomogeneous systems of ODE's.

Nancy Waller *and the proposer* give the following explicit solution by using the Cayley-Hanilton theorem:

$$X(t) = X(1) + \sum_{k=1}^{n} \frac{A(A-I)(A-2I)\cdots[A-(k-1)I]X(1)(t-1)^k}{k!}$$

A. S. Fernandez (E. T. S. Ingenieros Industriales de Madrid) uses the Lagrange interpolating polynomial to obtain

$$X(t) = \sum_{i=1}^{n} \frac{(-1)^{i-1} t^i}{(n-i)!(i-1)!} \prod_{j \neq i} (A - jI) X(1).$$

J. Roppert (Wirtschafts Universität Wien) also gives a solution if A has multiple eigenvalues by means of Jordan decomposition. Z. J. Kabala and I. P. E. Kinnmark (Princeton University) show how to solve (1) as above as well as when A has multiple eigenvalues. M. Latina (Community College of Rhode Island) in his solution notes that the method can be extended to more general Euler-Cauchy systems

$$\sum_{k=1}^{m} t^k A_k X^{(k)}(t) = 0.$$

Also solved by P. W. Bates (Texas A & M), G. A. Bécus (University of Cincinnati), J. Bélair (Université de Montreál), S. Coble (Student, University of Washington), C. Georghiou (University of Patras, Greece), O.

14　命题五则

Hajek (Case Western Reserve University), A. A. Jagers (Technische Hogeschool Twente, the Netherlands), D. James (San Antonio, Texas), R. A. Johns (University of South Carolina-Spartanburg), I. N. Katz (Washington University, St. Louis), G. Lewis (Michigan Technological University), H. M. Mahmoud (George Washington University), H. J. Oser (National Bureau of Standards), D. W. Quinn (Air Force Institute of Technology), P. H. Schidt (University of Akron), H. Türke (Universität Tübingen, FRG), G. C. Wake (Victoria University, New Zealand), two other solutions by Nancy Waller (Portland State University), J. A. Wilson (Iowa State University), P. T. L. M. van Woerkom (Bussum, the Netherlands) and one other solution by the proposer.

15

笛卡儿之梦

1981 年，Philip J. Davis 与 Reuben Hersh 合著的《数学经验》在美国出版之后，在国内外引起了强烈反响. 1983 年，该书荣获美国图书奖，后来被翻译成多种文字. 1982 年，我在美国 Brown 大学当访问学者，主人正是 Davis 教授，他把这本新书赠送给我，亲笔写上"为了友谊". Davis 希望我回中国后将书翻译成中文出版. 限于能力，我实在不能做到，于是谢绝了他的请求. 后来，Davis 如愿以偿，江苏教育出版社于 1991 年出版了该书的中文译本，译者是王前、俞晓群先生.

五年之后，同是这两位作者，又出版了《笛卡儿之梦》一书. 按照他们的说法，《数学经验》着眼于专业的观点，从数学的内部来看待数学，向读者描绘了所谓"做数学"的广阔图景. 而《笛卡儿之梦》这本书，则是从外部走向数学，同时也是从数学来看世界，向读者展示的是当数学应用于自然界和人类社会活动时所产生的影响. 从这些意义上来说，《数学经验》与《笛卡儿之梦》堪称"姊妹篇".

七年的时间，在瞬息中逝去了. 1989 年 7 月，我来到美国的犹他州杨伯翰大学土木工程系当访问学者，主人是 Sederberg. 而 Barnhill 和 Farin 早已离开 Utah 大学，搬到亚利桑那州立大学搞他们的 CAGD，据说 Barnhill 当上了相当于副校长的职位. 1989 年 11 月，一个 CAGD 国际会议在 Tempe 举行，主持人是 Barnhill 和 Farin，Sederberg、他的研究生和我也一同参加了会议. 在会上，我见到了 Davis 教授，时隔七年，他还是非常硬朗，像众星捧月一样，光艳夺目. 在 Tempe 的一家酒店里，Davis 的朋友和学生为他举行了一个特殊的"庆祝晚宴"，请帖上写着："荣誉归于 Philip Davis, 感谢他对于科学特别是对于计算机辅助几何设计的众多贡献. 1989 年 11 月 5 日（星期日）晚 7 点."参加晚宴的有 40 多人，大多是来自各国 CAGD 界的知名学者. 我的座位离 Davis 不远，他向我提起《笛卡儿之梦》这一新作，希望将它译成中文出版. 他还在我的请帖的背面，写上了书名和出版商的名字，至今我还保留着这一张珍贵的请帖.

我一直把 Davis 看做自己的良师益友，我感到有责任来满足他的要求. 1991 年 5 月我回到国内，当我准备开始着手这一项工作时，我感到由于我自己的英语水平和其他知识的限制，实在无法独立地翻译. 因为这一本书不但包含数学，还包含着其他学科：经济、哲学、历史、艺术、宗教、军事……几乎是古今上下无所不包，更不用说从文字角度来看，它还涉及古英语、外来语、诗歌、俚语、双关语、俏皮话……于是我想到了中国科大外语系周炳兰女士，请她来共同承担这项艰巨的任务. 两年的时间，非常艰苦，也合作愉快.

感谢台湾九章出版社的负责人孙文先先生，《笛卡儿之梦》终于在台湾地区以繁体字出版.

16

Over and Over Again

  我与 Sederberg 教授相识是在 1984 年 11 月，我们都在联邦德国 Oberwolfach 数学所参加"CAGD 中的曲面方法"会议. 他带着他的夫人 Branda 一起参加. Sederberg 个子高高的，长得很帅，Branda 也很漂亮，两个人都很年轻.

  他所在的学校叫做 Brigham Young University（杨伯翰大学），简称 BYU，城市叫 Provo，离盐湖城以南 1 小时的车程. BYU 是摩门教会的学校.

  之后四年多的时间，我们两人没有什么联系. 1989 年，我去信给他，有意访问 BYU，他欣然同意. 此前，他从没有接待过中国的访问学者，我是第一人. 后来，中国的学者多了起来，大多是我引荐给 Sederberg 的. 他多次对我说，其源头是我. 到他们系里的访问学者（长期或短期）有：齐东旭、王国瑾、汪国昭、郑健民、冯玉瑜、陈发来、李新等人.

  国际交流学者总有繁琐、文件式的事情. 有一天晚上，我接到美国的越洋电话，不是 Sederberg 打来的，而是 BYU 的某一位男士打来的. 他告诉我说，按照 BYU 的规定，学校不准抽烟，也不能喝咖啡和茶. 话讲得很客气，其他没有讲什么. 我早就知道 BYU 的规定，按规定办事就是了，足见学校非常重视.

  Sederberg 是美国总统青年奖得主，是非常优秀的人才. 到 BYU 的时候，我旁听了他的 CAGD 课程. Barnhill 的课程与 Sederberg 的截然不同，自己怎么感兴趣就怎么讲. Sederberg 讲代数曲线、隐式表示、结式, 而 Barnhill 这些都不讲.

从 1989 年 7 月开始, 访问共计 20 个月, 我们共同发表两篇文章. 我觉得从我这一方面来说, 努力不够. 数学证明是我做的, 但通篇的主旨, 我并不清楚. 他是学工程的 (尽管他的数学基础很好), 而我是职业数学工作者, 存在彼此交融的问题.

几年以前, 我曾经有过写一本书在美国出版的想法, 内容就是"迭代"(iteration), 已经有了很丰富的材料, 从初等数学到高等数学, 从有限次的迭代 (如: Douglas 和 Neumann 定理) 到无限次迭代 (极限过程). 我写信给"美国数学协会"(MAA), 告诉他们我这本书有什么特点, 有初等的数学竞赛的迭代, 还涉及高等数学的迭代, 包括 Bézier 曲线和曲面. "数学协会"的专家对我的想法很感兴趣, 但是, 要求我必须写几章当作"样版", 看看英文合不合格. 结果是通过了, 但是经过了很长的时间. 我向 Sederberg 征求合作写书的事情, 他同意了, 但是他的科研、教学和教会的事情很多, 不能抽出很多的时间. 我完全理解他的忙碌, 尽量减轻他的负担. 在 BYU, 我学会了用 LaTeX 打数学文章. 他负责我的英文, 他的英文打字速度非常快, 修改我的文章完全能够做到确切和优雅, 是我自己不可比拟的.

我们共同的书叫《Over and Over Again》, 是 Davis 教授为我们取的, 我觉得这个名字取得太妙了. 在美国, Over and Over Again 是一个习语, 人人都念. Iteration 是数学名词, 白话称为"一遍又一遍".

我们的书收录于 MAA 的"新数学文库", 1961 年开始创立的. 1997 年, 我们的书第一次印刷, 序号是 39, 是科普性的读物. 原是为了美国中学生能看懂, 实际上有很多的内容是大学课程. "文库"作者当中, 有很多美国的著名数学家.

我们非常感谢主编和编辑们为此书的出版付出的辛勤劳动, 特别是编辑 Peter Ungar 教授和顾问编辑 Anneli Lax 教授. Ungar 教授认真阅读了全书, 对 Bézier 相应章节给出了收敛性证明, 扩充了第 29 章的内容, 写出了变差缩减矩阵和整个附录. Ungar 教授和作者们保持着长期的往来联系. Anneli 负责与其他的编辑协调, 对改进本书的陈述风格起了很大的作用. 她亲自认真地看校样. 她的丈夫是著名数学家 Peter Lax, 他的关于圆的等周问题的新证明, 是 Anneli 推荐给作者们的. 1999 年, Anneli 罹患胰腺癌不幸逝世. 欣慰的是, Anneli 看到了我们的书出版问世.

  Sederberg 和我应当感谢我们共同的朋友齐东旭教授. 在写作我们的书的时候，Sederberg 邀请齐东旭访问 BYU. 这本书的第 29 章 "移动平均"，就是根据他的初稿写成的. 李岳生、齐东旭合著的专著《样条函数》（科学出版社，1979），包含了他们的研究成果. 记起有一件趣事，齐东旭只是在 BYU 访问半年，他突然心血来潮，想买一辆旧车玩玩. 从买车、进驾驶学校、陪练、考驾照，总归要花不少时间，常人不敢想的他顺利办成了. 他的二手车（也可能是 N 手车）为他辛劳服务，从不抛锚. 离开 BYU 前，他转手卖掉旧车，还没有亏本. 我多少次做他的陪练，就是说，没得到驾照前，他不能独立开车，必须有个有驾照的人在他的身旁，陪他上路. 我发明了一个 "判别法"，如果左脚脚心绷紧，说明你还是很紧张. 齐东旭说，这个办法很灵.

  我感谢 Peter Ungar 教授、Anneli Lax 教授和其他编辑们，也特别感谢 Tom Sederberg 教授，没有他们的帮助和支持，就没有这本书. 1997 年本书开始发行，后来作为 "美国数学协会" 畅销书之一，直到今天仍在美国有售.

  Tom 喜欢讲笑话，他说："有一位男士清早去上班，在街区转角处，每天都会碰到一个卖糖饼的老太太，在寒风中兜售糖饼. 男士见她可怜，每天给她五毛钱，不要她的糖饼. 有一天，老太太对男士说：'今天涨价了，糖饼一块钱一个.'"

17

数学分析教程

我在天津南开大学当学生的时候，教我们"数学分析"的是周学光教授，教了两年的时光，教材是苏联数学家辛钦写的《数学分析简明教程》. 1958 年，我来到北京中国科学技术大学，担任关肇直教授的助教，辅导学生的"数学分析"，两个年级总共三年. 1978 年，我第一次主讲"数学分析"，仅过了一年，就不能再教了，因为我要到美国进修，学校要办外语培训班. 90 年代，我又开始讲"数学分析"，给数学系和少年班上课. 我积累了很多关于数学分析的材料，以及对多年来教学相长的体会，还有我自己的心得. 我曾经戏言，各种数学分析教材与牙膏品牌一样多，大同小异，质量上乘者很少.

1985 年，中国科大教材《数学分析》在北京高等教育出版社出版，作者是何琛教授等三人. 我对此书的第一印象是，多变量微分学采取向量、矩阵记号，为一大特色，二重积分也有新意. 这套数学分析教材，已经在中国科大数学教学中起了重要的作用，在全国同类书籍中产生了积极的影响.

多年来，我一直有编写一本数学分析教材的冲动，主旨是"删繁就简，标新立异". 数学杂志上一些精彩的证明我都记录下来，备课时一鳞半爪的心得也不放过. 80 年代我学习的 CAGD，移植过来就是数学分析取之不尽的材料.

如何选择合作者？当然是史济怀为首选. 在《数学分析》一书中，他是第二作者，在本书第三册中，他是主笔，我知道他的想法，他们的书我也教过多遍. 他欣然同意我们合作的打算. 实际上，我们配合默契，直来直去，毫不隐瞒，合作愉快. 我与他都是 1958 年来到中国科技大学的，他是 9 月到的，而我晚了一个月.

我们的《数学分析教程》(简称《教程》) 第一册，由江苏教育出版社 1998 年 10 月出版. 随后，由出版社的推荐，台湾凡异出版社用繁体字将我们的《教程》出版 (2001 年). 后因江苏教育出版社不再出版大学教材，改为北京高等教育出版社出版，成为普通高等教育"十五"国家级规划教材 (2003 年)，重印 10 次. 2012 年 8 月，再次改写成为第三个版本，由中国科学技术大学出版社出版. 我们的《教程》经历了 4 个出版社，总共重印 16 次.

我们的《教程》到底有什么特点呢？归结起来就是：

(1) 在许多地方添加了新的材料，证明了"Riemann 可积的充分必要条件是被积函数在积分区间上的不连续点的集合是一零测集"，不需要实变函数理论，只需数学分析技巧. 有了这一定理，就可以删除关于可积性的许多讨论，从总体上来看缩短了篇幅. 增加了二元凸函数的理论和应用，证明了"三角域 Bézier 网如果是凸的，那么它对应的 Bernstein 三角曲面也是凸的". 采用了 Peter Lax 对圆的等周性质的优美证明. 收录了能充满整个正方形的 Schoenberg 连续曲线.

(2) 在第二章"函数的连续性"的最后，我们介绍了"混沌现象"，叙述并证明了李天岩和 Yorke 的《周期 3 蕴涵混沌》的著名定理 (1975).

(3) Bernstein 多项式 (1912) 是国内外数学分析必读内容，主要用来在闭区间上来一致逼近连续函数，虽然它是构造性的，但是收敛的速度非常缓慢. 1960 年前后，法国工程师 Bézier 创造的、后来被人们所称的 Bézier 曲线和曲面，被成功地运用到汽车设计之中，已成了当今 CAGD 和 CG (计算机图形学) 的理论基础. 人们发现，所谓的 Bézier 曲线 (曲面) 只不过是向量值形式的一元 (二元) Bernstein 多项式. Bézier 的成功之处是它充分利用了 Bernstein 多项式的"保形性"，这恰好是传统的数学分析关于 Bernstein 多项式不曾谈到过的.

Bernstein 多项式的"磨光性质"在 Kelisky-Rivlin 的定理 (《Pacific J. of Math.》, 1967) 中得到了充分体现. 原先，他们的证明里要用到矩阵特征值和特征向量，而我们的简化证明可以使中学生都能看得懂.

(4) 在用 Van der Waerden 方法构造处处连续而处处不可微的函数之后，介绍了"分形几何"的大意. 传统的数学分析只是把这个例子当成一个"反例",而我们试图告诉学生,在自然界和社会的现象中,到处存在着这种不规则、不光滑的东西.

在构造 Van der Waerden 函数时,既是非常严格的,同时又免去了传统证明中一系列琐碎的区间分段.

(5) 在写作风格上,我们很不赞成一些数学书中的所谓"标准写法",那些语言像一串电码,没有任何感情色彩.我们力图把读者当成朋友,平等对话,娓娓谈心.

1998 年 10 月,我们的新《教程》发给学生了,我们非常激动.我们两人轮流教"数学分析",要么是数学系,要么是少年班,两人轮流"坐庄".

不过,好景不长,只不过是两年多的时光,一片阴霾开始笼罩在我的头顶.那是 2000 年 12 月,我在科大数学系"四牌楼"一层教室上"数学分析",对象是 1998 级数学系学生,在上课当中,我突然讲不出话来,我对学生说:"我想休息一下,等一会儿再讲."学生鸦雀无声,只有我知道,那叫做"失语",空 5 分钟、10 分钟,是可以缓过来的.我已经搞过三次了,过去不以为意,今天在课堂讲课时出现,不能不重视了,我必须到安医(安徽医科大学第一附属医院)做一个全面的检查.

我的老毛病是高血压病和冠心病,做了心电图和彩超心电图以后,没有发现严重的不适,最后只剩下脑 CT. 第一次做脑 CT,庞大的机器使我有些紧张.快到中午了,只剩下我一个病人,我对医生说:

"我没有问题吧?"

"你有问题,两点钟上班后,再做一个加强的脑 CT,带你的爱人来."

"我爱人不在合肥."

"那就请你领导来."

医生吃中午饭去了,而我没心思吃饭,一下子想到脑癌,简直是世界末日快要降临了.我打电话给黄素琴同志,央求她两点钟陪我到安医,权当我的"领导".医生与"领导"谈话时,我这个病人不能在场,我更紧张了.

黄素琴从医生办公室出来,对我说:"没有事,是良性的,叫做脑膜瘤,像蛋黄一样大,要开刀的啊!"我只知道有"脑膜炎",脑膜瘤我是第一次听到.

12 月下旬,正好学期结束,学生停课了.我到北京接受脑膜瘤手术,在北京航天总医院,北京天坛医院脑外科专家杨骏为我主刀,据说手术进行不到一个小时,一切顺利.在 ICU 待了 5 天,我自己是糊里糊涂的.我想喝水时，居然讲不出话来，我五只指头做成圆形，往嘴里一放——我想喝水，护士全都明白. 在开刀之前，有专家对我说，由于你的部位开刀时可能触碰语言神经，可能很长时间语言讲得不利索，这是事先知道的有这种可能，但性命要紧，其他都无所谓了. 从 2001 年往后，我不得不告别讲台，留下了永远的遗憾.

自我们的《教程》于 1998 年问世以来，史济怀一直在主讲"数学分析". 虽然他年过 80 岁，但仍然身体硬朗、腰腿犹轻，上楼梯"两步一走". 学校教务处为他制作了课堂即时录像，相当完整，供读者观看. 在他教学的过程中，他曾经修改两次，主要是将涉及 CAGD 过多的内容删去，并增加了其他的内容. 虽然我对 CAGD 有所偏爱，在教学实践中，我也逐渐同意他的说法. 史济怀教授是国家级名师，肯定增加了我们《教程》的知名度.

在互联网中，我们的《教程》好评如潮，备受称赞，我们作者很受感动. 但愿人长久，我们会不断地改进，不辜负读者的厚爱.

18

重访 BYU

2008 年 5 月间，Sederberg 教授发来电子邮件，问我有没有兴趣访问 BYU，他当上了 BYU 理学院的副院长，分管数学系. 数学系有几个学生天资聪颖，准备今年冬季参加 Putnam 数学竞赛，他要我来 BYU，介绍我做题和命题的经验，他知道我搞过 IMO. 我斩钉截铁地告诉他，我不行，我英语听不懂.

他回答说："访问一天，行不行？"

我说："也不行."

他说："有人给你当翻译，怎么样？"

我说："那行！"

绕来绕去，我终于明白了，他们的 Putnam 数学竞赛指导教师是美籍中国教授欧阳天成，尽管我们没有见过面，但我心里踏实多了. 通过几番电话和通信，彼此渐渐地走近了. 我们两个人都是南开数学系出身的，我大他 14 岁，他的父亲是数学系的教授，但我没有听过他的课. 我告诉欧阳，我不能为学生讲课，想问题很慢，但我有一些经验，可以当你的参谋，出出主意. 我把有脑膜瘤后遗症的事情，一五一十告诉他，他表示理解. 连续五年，欧阳教授盛情邀请我参加 BYU 的数学竞赛，每年一个月左右. 只是因为我身体体力下降，才恋恋不舍地离开杨伯翰大学.

到达盐湖城机场那一天，欧阳和他的博士研究生严夺魁在机场接我. 虽然我们没有见过面，但是一见如故.

美国 Putnam 数学竞赛，是为了纪念数学家 William Lowell Putnam，始于 1938 年，至今办了 70 多届，是世界上享有盛名的数学竞赛之一. 参加者为北美的大学生，主要是美国和加拿大的大学生，内容包括数学分析、线性代数、组合学、概率论等等，题材广泛，也不靠任何高深数学知识，连中学生都可以一试. 历届竞赛试题，与时俱进，难度逐渐提高. 与 Putnam 竞赛的题目、解答、技巧和方法相关的图书，多种多样，在全世界畅销.

欧阳教授主讲的数学竞赛，我场场必到，家庭作业或模拟考试由欧阳做主，我是从旁提一些建议. 有些学生果然进步很大，Putnam 竞赛成绩优胜者 BYU 榜上有名，过去只是梦想.

数学系长年租了一间房间，一室一厅带卫生间、厨房，供访问学者使用，我就在这公寓里. 严夺魁经常带我去买菜，给我很多的帮助. 他学成回国之后，在北京找到了教职.

Tom 依然非常忙碌，教学、科研、行政、教会、家庭，都是很大的负担，他的夫人 Branda 前几年得了一种可怕的疾病，叫做"硬皮病"，是免疫功能出了毛病，身体一天一天地消耗. Tom 每天陪伴在她的身边，他不能够去作报告，更不能出国讲学，只能够去当天打来回的地方. Tom 对 Branda 不离不弃，相亲相爱，我们为之感动.

每一次到 BYU 来，Tom 总要请我吃两次饭，一次到他们家里来，见见 Branda 和他们的子女，另一次在饭馆，只有我们两个人在场. 每年我从加州到 BYU 来，总要带一些上海的"大白兔奶糖"送给他们的孩子. 这里有一段故事：1992 年，我邀请 Tom 到中国科大讲学，Branda 和大儿子、大女儿同行. 第一站就是复旦大学，两个孩子喜欢吃上海大白兔奶糖. 然后到浙江大学，在杭州，惊悉我的岳父在北京病逝，享年 88 岁，而我不能赶回去奔丧. 我们一行 5 人，乘运 -7 飞机从杭州到达合肥，开始他们对中国科大的访问. Tom 第一次见到陈发来，在学术交流中留下了深刻的印象，后来他们两人成为论文合作者和互相往来的好朋友. 中国之行的最后一站是北京，Tom 一家由齐东旭教授接待.

我最后一次访问 BYU 是 2012 年 9 月至 10 月，正好严夺魁访问他的老师欧阳教授，时间过得很快，5 年不见面了. 我们都住在数学系租来的公寓里，我睡卧房，他睡客厅，晚上支开折叠床，相当费事，他尽力照顾我. 白天我们共用一个办公室，晚餐由小严料理.

18 重访 BYU

2012年10月初,Tom发来电子邮件,他的夫人Branda去世了,她患病十几年,每况愈下,全靠Tom照顾和服侍,离去也是一种解脱.摩门教会重视家庭,认为人死之后可以在天国同家人们在一起.举行告别仪式的教堂,离欧阳教授家非常近,欧阳让我住在他的家里,免得手忙脚乱.欧阳和我一起到教堂,这时早上8点整,我们是第一批宾客,Tom西装革履站在Branda的棺材前,我们向Branda三鞠躬,可以看到她闭着眼睛,神态安详,满头白发.

Tom对我们微笑着说:"Branda漂亮吗?"

我们说:"她很漂亮."

常庚哲简历

1　个人信息

性别　　男
出生年月　1936 年 8 月 17 日
退休时间　2001 年 8 月
出生地　湖南省长沙市
学历　　1954 年湖南省长沙市第一中学高中毕业
　　　　　1958 年天津市南开大学数学系本科毕业
曾任职　中国科学技术大学数学系教授（1985~2001）
　　　　　博士生导师（1990~2001）

地址　　安徽省合肥市
邮编　　230026
电话　　0551-63601018
Email　changgz@ustc.edu.cn

专长　　数值分析
　　　　　计算机辅助几何设计
　　　　　应用逼近理论

2　获奖情况

1978~1979 年中国科学院重大科技成果三等奖
1992 年中国科学院自然科学二等奖
1995 年宝钢教育基金"优秀教师特等奖"
2007 年中国工业与应用数学会几何设计与计算贡献奖
2008 年中国科学技术大学"孺子牛"特殊贡献奖
2009 年中国科学技术大学"困学守望"终身成就奖

3　学术任职

学术团体　安徽省数学会理事长（1997~2002）
　　　　　　美国《数学评论》评论员
学术刊物　国际刊物《Computer Aided Geometric Design》编委
　　　　　　（1984~1999）
　　　　　　《高等学校应用数学学报》编委
　　　　　　《数值计算与计算机应用》编委
　　　　　　《计算机辅助设计与图形学学报》编委
　　　　　　《工程数学学报》编委
　　　　　　《教育与现代化》编委
　　　　　　《高等数学研究》特邀编委

4　国际学术交流

1980 年 8 月~1981 年 12 月	美国 Utah 大学数学系访问学者
1982 年 1 月~7 月	美国 Brown 大学应用数学系访问学者
1982 年 8 月	加拿大 Wilfred Laurier 大学数学系访问学者
1984 年 11 月~1985 年 1 月	联邦德国 Darmstadt 工业大学数学系访问学者并顺访柏林工业大学
1985 年 6 月~9 月	意大利国际理论物理中心（ICTP）访问学者

1987 年 2 月~4 月	联邦德国 Darmstadt 工业大学、Duisburg 大学、柏林工业大学访问学者
1988 年 2 月~5 月	意大利 ICTP 访问学者并顺访南斯拉夫的 Zagreb 大学
1988 年 5 月	瑞典 Linköping 大学数学系访问学者
1988 年 7 月	作为出席国际数学奥林匹克的中国代表队的领队,前往澳大利亚首都堪培拉
1989 年 7 月~1991 年 5 月	美国杨伯翰大学访问学者
1992 年 11 月~12 月	瑞典 Linköping 大学数学系访问学者
1993 年 10 月~1994 年 3 月	美国杨伯翰大学访问学者
1994 年 7 月	马来西亚 Sains 大学数学系访问学者并顺访新加坡国立大学数学系
1995 年 6 月~7 月	新加坡国立大学数学系访问学者
2008 年 8 月~9 月	美国杨伯翰大学访问学者
2009 年 3 月	美国杨伯翰大学访问学者
2010 年 10 月~11 月	美国杨伯翰大学访问学者
2011 年 10 月~11 月	美国杨伯翰大学访问学者
2012 年 09 月~10 月	美国杨伯翰大学访问学者

5　出版著作

译书

1. Su B Q, Liu D Y. Computational Geometry-Curve and Surface Modeling[M]. Boston: Academic Press,1990.
2. Davis P J, Hersh R. 笛卡儿之梦[M]. 常庚哲,周炳兰,译. 台北:九章出版社,1997.

著书

1. 常庚哲,伍润生. 复数与几何[M]. 北京:人民教育出版社,1963.
2. 常庚哲,伍润生. 复数与几何[M]. 香港:香港智能教育出版社,2003.

3. 常庚哲，许有信. 飞机外形计算的数学基础[M]. 北京：国防工业出版社，1977.

4. 常庚哲. 抽屉原则及其他[M]. 上海：上海教育出版社，1978.

5. 常庚哲. 抽屉原则及其他[M]. 台北：凡异出版社，1994.

6. 常庚哲. 复数计算与几何证题[M]. 上海：上海教育出版社，1980.

7. 常庚哲，苏淳. 奇数和偶数[M]. 上海：上海教育出版社，1986.

8. 常庚哲，苏淳. 奇数和偶数[M]. 台北：凡异出版社，1994.

9. 常庚哲. 初中数学竞赛妙题巧解[M]. 上海：上海科学技术出版社，1988.

10. 常庚哲. 变换与数学竞赛[M]. 北京：中国少年儿童出版社，1993.

11. 常庚哲. 曲面的数学[M]. 长沙：湖南教育出版社，1995.

12. 常庚哲. 神奇的复数[M]. 台北：九章出版社，1996.

13. Chang G Z, Sederberg T W. Over and Over Again[M]// 美国数学协会新数学文库，1997.

14. 常庚哲，史济怀. 数学分析教程：第一册[M]. 南京：江苏教育出版社，1998.

15. 常庚哲，史济怀. 数学分析教程：第二册[M]. 南京：江苏教育出版社，1999.

16. 常庚哲，史济怀. 数学分析教程：第三册[M]. 南京：江苏教育出版社，1999.

17. 常庚哲，史济怀. 数学分析教程：第一册[M]. 台北：凡异出版社，2001.

18. 常庚哲，史济怀. 数学分析教程：第二册[M]. 台北：凡异出版社，2001.

19. 常庚哲，史济怀. 数学分析教程：第三册[M]. 台北：凡异出版社，2001.

20. 常庚哲，史济怀. 数学分析教程：上册[M]. 北京：高等教育出版社，2003.

21. 常庚哲，史济怀. 数学分析教程：下册[M]. 北京：高等教育出版社，2003.

22. 常庚哲，史济怀. 数学分析教程：上册[M]. 合肥：中国科学技术大学出版社，2012.

23. 常庚哲，史济怀. 数学分析教程：下册[M]. 合肥：中国科学技术大学出版社，2013.

24. 常庚哲. 抽屉原则[M]. 合肥：中国科学技术大学出版社，2012.

25. 常庚哲. 磨光变换[M]. 合肥：中国科学技术大学出版社，2013.

6 研究论文和综合报告

1. 常庚哲. Coons 曲面介绍[J]. 计算机应用与应用数学，1977，12：1-24.

2. 常庚哲，吴骏恒. 贝齐尔曲线、曲面的数学基础及其计算[J]. 国外航空，1979（1-6）.

3. 常庚哲. 三次样条函数的两点注记[J]. 数学的实践与认识，1979（2）：55-64.

4. 常庚哲，吴骏恒. 关于 Bézier 方法的数学基础[J]. 计算数学，1980（1）：41-49.

5. Chang G Z, Wu J H. Mathematical Foundations of Bézier Technique[J]. Computer Aided Design, 1981, 13：133-136.

6. Chang G Z. A Proof of a Theorem of Douglas and Neumann by Circulant Matrices[J]. Houston J. of Math., 1982（8）：15-18.

7. Chang G Z. Provinng Pedoe's Inequality by Complex Number Computation[J]. Amer. Math. Monthly, 1982, 89：692.

8. Chang G Z. Matrix Formulations of Bézier Technique[J]. Computer-Aided Design, 1982, 14：345-350.

9. Chang G Z, Shang Z. A Simple Proof for a Theorem of Kelisky and Rivlin[J]. 数学研究与评论，1983（3）：145-146.

10. Chang G Z. Generalized Bernstein-Bézier Polynomials[J]. J. Comp. Math., 1983（1）：322-327.

11. Chang G Z, Feng Y Y. Error Bound for Bernstein-Bézier Triangular Approximation[J]. J. Comp. Math., 1983（1）：335-340.

12. Chang G Z, Davis P J. A Circulant Formulation for the Napolean-Douglas-Neumann Theorem[J]. Linear Algebra and its Application, 1983, 54: 87-95.

13. Chang G Z. Subdivision of the Bézier Curve[J]. Int. J. Number Methods in Engineering, 1983, 19: 1227-1233.

14. Chang G Z, Davis P J. Iterative Processes in Elementary Geometry[J]. Amer. Math. Monthly, 1983, 90: 421-431.

15. Chang G Z, Davis P J. The Convexity of Bernstein Polynomials over Triangles[J]. J. Approx. Theory, 1984, 40: 11-28.

16. Chang G Z. An Elementary Proof of the Convergence for the Generalized Bernstein-Bézier Polynomials[J]. J. Comp. Math., 1984 (2): 84-87.

17. Chang G Z. Bernstein Polynomials via the Shifting Operator[J]. Amer. Math. Monthly, 1984, 91: 634-638.

18. 常庚哲. 用 Bézier 函数证明 Bernstein 逼近定理[J]. 数学年刊: A 辑, 1984 (5): 321-324.

19. 常庚哲, 冯玉瑜. 计算几何与 Bernstein 多项式[J]. 应用数学与计算数学, 1984 (3): 1-10.

20. 常庚哲, 冯玉瑜. 三角域上的 Bernstein 多项式的凸性[J]. 自然杂志, 1984 (7): 779-800.

21. 常庚哲, 冯玉瑜. 样条变差减缩算子迭代极限的一个简单证明[J]. 系统科学与数学, 1984 (4): 165-172.

22. 常庚哲. 三次样条函数连续性方程的新推导[J]. 数学的实践与认识, 1984 (4): 58-61.

23. 常庚哲, 冯玉瑜. 定义在矩形域上的 Bernstein 多项式的迭代极限[J]. 工程数学学报, 1984 (1): 137-141.

24. Chang G Z, Feng Y Y. An Improved Condition for the Convexity of Bernstein-Bézier Surfaces over Triangles[J]. Computer Aided Geometric Design, 1984 (1): 279-283.

25. Chang G Z, Feng Y Y. A New Proof for the Convexity of the Bernstein-Bézier Surface over Triangles[J]. 数学年刊: B 辑, 1985（6）: 171-176.

26. Chang G Z. Planar Metric Inequalities Derived from Vandermonde Determinants[J]. Amer. Math. Monthly, 1985, 92: 495-499.

27. Chang G Z, Su B Q. Families of Adjoint Patches for a Bézier Triangles Surfaces[J]. Computer Aided Geometric Design, 1985（2）: 37-42.

28. Chang G Z, Hoschek J. Variation Diminishing Properties and Convexity for Bernstein Polynomials over Triangles[J]. Multivariate Approximation Theory Ⅲ: Birkhäuser Verlag, 1985: 61-70.

29. Chang G Z, Feng Y Y. A Pair of Compatible Variations for Bernstein Triangular Polynomials[J]. 逼近论及其应用, 1989（5）: 1-10.

30. 常庚哲. 数学中的磨光变换[J]. 自然杂志, 1985（8）: 830-836.

31. Chang G Z. Limit of Iterates for Bernstein Polynomials Defined on Higher Dimensional Domains[J]. 科学通报, 1986, 31: 157-160.

32. Chang G Z, Hoschek J. Convergence of Bézier Triangular Nets and a Theorem of Pólya[J]. J. Approx. Theory, 1989, 58: 247-258.

33. 常庚哲，张景中，杨路. 高维单形上 Bernstein 多项式的凸性定理的逆定理[J]. 中国科学: A 辑, 1989（6）: 588-599.

34. Chang G Z, Zhang J Z. Converse Theorem of Convexity for Bernstein Polynomials over Triangles[J]. J. Approx. Theory, 1990, 61: 265-278.

35. 常庚哲，陈发来. 单纯形上 Bernstein 多项式的凸性定理的逆定理的一个简短证明[J]. 数学的研究与评论, 1991（11）: 275-277.

36. Chang G Z, Sederberg T W. Best Linear Common Divisors for Approximate Degree Reduction[J]. Computer Aided Geometric Design, 1993, 25: 163-168.

37. Chang G Z, Sederberg T W. Isolator Polynomials[J]. Algebra Geometry and its Applications: Springer-Verlag, 1994: 507-512.

38. Chang G Z, Sederberg T W. Nonnegative Quadratic Bézier Triangular Pathces[J]. Computer Aided Geometric Design, 1994（11）: 113-116.

39. Chang G Z, Andersson L E, Elfving T. Criteria for Copositive Matrices Using Simplices and Barycentric Coordinates[J]. Linear Algebra and its Applications，1995, 220：9-30.

40. 常庚哲, 蒋继发. 用部分积分法求解常系数高阶非齐次常微分方程 [J]. 大学数学，2003, 19: 76-79.

41. 常庚哲. 一个带有 Γ 函数的二重积分[J]. 高等数学研究，2011（2）：1-3.

42. 常庚哲. 单变量函数的积分形成的迭代[J]. 高等数学研究，2012（1）：27-29.

43. Yan D K, Liu R C, Chang G Z. A type of Multiple Integral with Log-gamma Function[EB/J]. http://www.docin.com/p-927424404./html.

7 在国外数学期刊上提出的问题

1125	Mathematics Magazine (U.S.A)
257	Two-year College Mathematics Journal (U.S.A)
6375	The American Mathematical Monthly
E3021	The American Mathematical Monthly
E3031	The American Mathematical Monthly
955	Crux Mathematicorum (Canda)
82-19	SIAM Review (U.S.A)
83-2	SIAM Review (U.S.A)
83-3	SIAM Review (U.S.A)
84-12	SIAM Review (U.S.A)
85-3	SIAM Review (U.S.A)
84-16	The Mathematical Intelligence
第二题	第 27 届国际数学奥林匹克竞赛（1986）